FEARON'S

Basic Mathematics

SECOND EDITION

Katherine P. Layton
Joan Kreisl

Globe Fearon Educational Publisher
Paramus, New Jersey

Paramount Publishing

Pacemaker Curriculum Advisor: Stephen C. Larsen
Stephen C. Larsen holds a B.S. and an M.S. in Speech Pathology from the University of Nebraska at Omaha, and an Ed.D. in Learning Disabilities from the University of Kansas. In the course of his career, Dr. Larsen has worked in the Teacher Corps on a Nebraska Indian Reservation, as a Fulbright senior lecturer in Portugal and Spain, and as a speech pathologist in the public schools. A full professor at the University of Texas at Austin, he has nearly twenty years' experience as a teacher trainer on the university level. He is the author of sixty journal articles, three textbooks, and six widely used standardized tests including the Test of Written Learning (TOWL) and the Test of Adolescent Language (TOAL).

Subject Area Consultant: Helen Butler Miller
Helen Butler Miller holds an M.S. in Mathematics from California State University, Hayward, California. She has taught various math courses as a fulltime instructor at Diablo Valley College, Pleasant Hill, California.

Second Edition Consultant: Ronn Yablun
Ronn Yablun holds a B.S. from Northern Illinois University. He is chairman of the math department at Northridge Junior High School in California. He is also the founder of Mathamazement, a learning center for remediation and enrichment in math.

Editor: Tony Napoli
Contributing Editor: Sharon Wheeler
Text Design: Jill Casty and Company
Cover Design: Marc Ong, Side by Side Studios

Photo Credits: San Diego Zoo 34; The Bettmann Archive 56, 262, 300; Indianapolis Motor Speedway, Photo by Ron McQueeney 76; National Archives 94, 180; Ron DiDonato Photography 128, 200; Shulee Ong and Ted Williams 142; John Herr 160; Comstock Stock Photography, Inc. 214; Sequoia Hospital District, Redweed City, CA 226; Chicago Tribune Company 244.

Additional Photos Courtesy of: Cleveland Indians 2; Los Angeles Public Library 18; The Pacific Stock Exchange, Inc. 112; Master Sergeant Ken Hammond 280; Boston Red Sox 310.

About the Cover Photograph: *© The Stock Market/Nick Koudis, 1990.* From the first time primitive people began counting objects with sticks, stones, or their ten digits (fingers), mathematics has been a way to help people understand and deal with their environment. Today we have calculators and computers to help us, but an understanding of basic math has to come first.

ISBN 0–8224–6898–0

Printed in the United States of America
3. 10 9 8 7 6 5 4 3
Cover Printer/NEBC
DO

Contents

Appendix

A Note to the Student

Today, students often wonder why they *still* have to study math. After all, calculators and computers are practically everywhere. These wonderful machines can answer the most complex problems in seconds.

Certainly there is some truth to that. But a calculator or a computer can only do what you tell it to do. You have to ask the right question to get the right answer. If you don't know whether to multiply or divide to solve a problem, a calculator can't help you. To decide what to do, you must know something about basic mathematics. And your calculator can't tell whether or not an answer makes sense. Only human judgment can do that. But you must know something about basic mathematics to make reasonable judgments.

You already use some form of mathematics in your life every day. You tell time. You measure. You spend money and count your change. You figure out how long it will take to get from one place to another. In fact, you couldn't survive in today's fast-paced world without using some basic mathematics skills.

The purpose of this book is to help you develop the math skills you need to succeed as well as to survive. Chapter by chapter you will learn about whole numbers and how to add, subtract, multiply, and divide them. You will also learn about fractions, decimals, percents, different systems of measurement, and other basics of mathematics. Some of this information may not seem very useful right now. But a solid understanding of basic math will help you make good decisions all your life—at school, at home, and on the job.

Throughout the book you'll find notes in the margins of the pages. These friendly notes are there to make you stop and think. Sometimes they comment on the material you are learning. Sometimes they remind you of something you already know. And sometimes they give you interesting bits of "math trivia."

You will also find several study aids in the book. At the beginning of every chapter, you'll find **Learning Objectives**. They will help you focus on the important points covered in the chapter. And you'll find **Words to Know**, a look ahead at the vocabulary you may find difficult. At the end of each chapter, a **Summary** will give you a quick review of what you've just learned.

Everyone who put this book together worked hard to make it useful, interesting, and enjoyable. The rest is up to you. We wish you well in your studies. Our success is in your accomplishment.

UNIT ONE

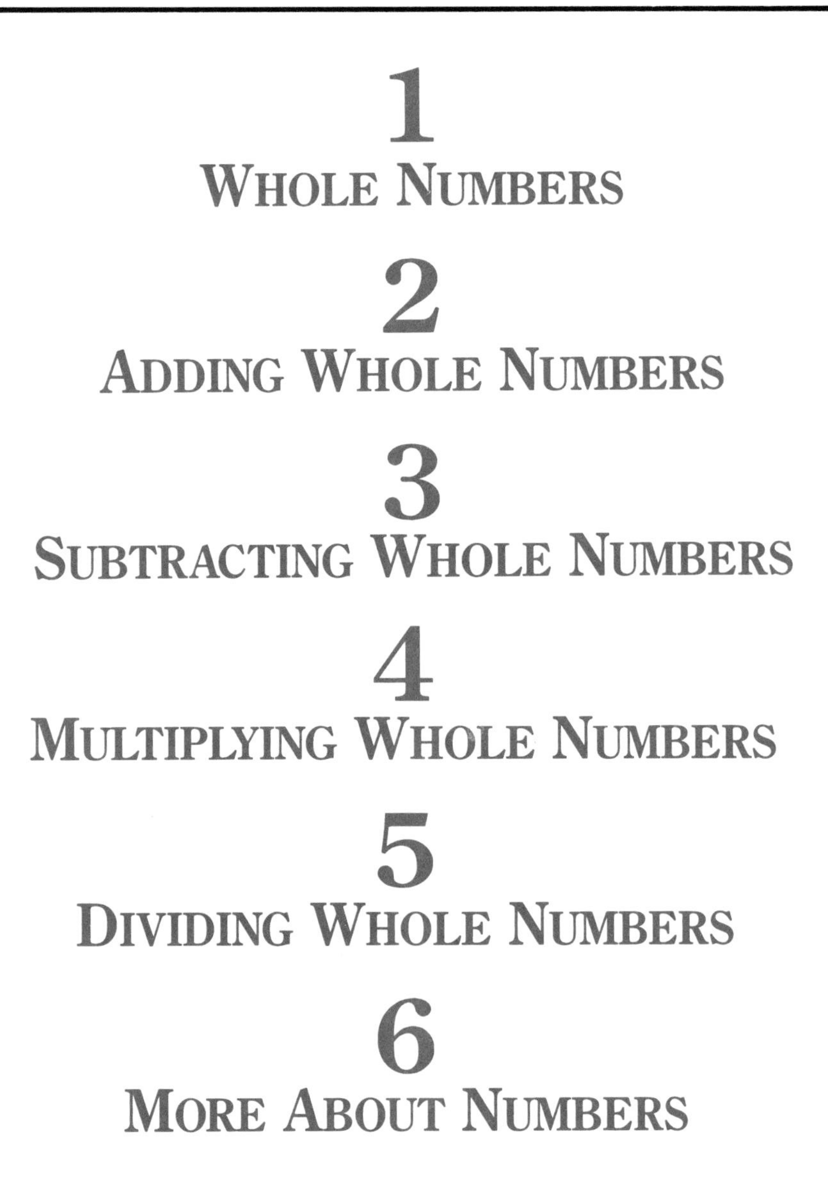

1
WHOLE NUMBERS

2
ADDING WHOLE NUMBERS

3
SUBTRACTING WHOLE NUMBERS

4
MULTIPLYING WHOLE NUMBERS

5
DIVIDING WHOLE NUMBERS

6
MORE ABOUT NUMBERS

WHOLE NUMBERS

1

This is a scoreboard from a major league baseball stadium. All the numbers used to keep score are whole numbers. In this chapter, you will learn all about whole numbers.

Chapter Learning Objectives

1. Read and write whole numbers
2. Count on a number line
3. Identify odd and even numbers
4. Recognize place value
5. Rename numbers
6. Round whole numbers

WORDS TO KNOW

■ whole number	0, 1, 2, 3, 4, 5, 6, 7, and so on
■ digit	the symbols used to write numbers: 0, 1, 2, 3, 4, 5, 6, 7, 8, and 9
■ round	to change a number to the nearest ten, hundred, thousand, and so on
■ number line	numbers in order shown as points on a line
■ rename	to break down a number into its parts; 28 can be renamed as 2 tens and 8 ones
■ even number	a number that ends in 0, 2, 4, 6, or 8, except zero
■ odd number	a number that ends in 1, 3, 5, 7, or 9

1.1 What Is a Whole Number?

The numbers 0, 1, 2, 3, 4, 5, and so on, are called **whole numbers.** Whole numbers are used to count. They tell how many or how much.

Examples:

I need 3 more dollars.

There are 16 people in this class.

A crowd of 6,972 people attended the basketball game.

The score was 82 to 76.

The population of Centerville is 2,346,721.

PRACTICE

Which of the following are whole numbers?

1. 27 **2.** 17.5 **3.** 0 **4.** $5\frac{1}{2}$

1.2 The Number Line

This is a **number line.** It shows numbers in order as points on a line. This number line shows the numbers from 0 to 16. A number line is read from left to right.

0 1 2 3 4 5 6 7 8 9 10 11 12 13 14 15 16

Order on the Number Line

The numbers on a number line get bigger from left to right. The last number on *this* number line is 16. But number lines do not end at 16. They can go on and on.

Number lines are read the same way we read sentences—from left to right.

PRACTICE

A. Choose the larger number in each pair. Write your answers on a separate sheet of paper.

1. 3, 7 **2.** 9, 5 **3.** 6, 11 **4.** 12, 15 **5.** 14, 4

B. Copy the number line on a separate sheet of paper. Fill in each blank with the correct number.

0 1 __ 3 4 __ 6 __ 8 9 __ 11 12 __ 14 __

C. The number line below goes from 25 to 42. Copy the number line on a separate sheet of paper. Fill in each blank with the correct number.

25 __ 27 __ 29 30 __ 32 __ 34 35 __ 37 __ 39 __ 41 42

D. There are twelve short number lines below. Copy them on a separate sheet of paper.

Fill in the blanks. The first one has been done for you.

1. 4 5 6 **4.** __ 10 __ **7.** __ 12 __ **10.** __ 14 __

2. __ 11 __ **5.** __ 16 __ **8.** __ 19 __ **11.** __ 30 __

3. __ 33 __ **6.** __ 45 __ **9.** __ 49 __ **12.** __ 60 __

1.3 Odd and Even Numbers

Look at the numbers line below. Notice the numbers in color. They are called **even numbers**. The numbers in black are called **odd numbers** (except zero).

Notice the order of odd and even numbers on the number line. The number 1 is an odd number. Next comes 2, an even number. Then comes 3, an odd number. The pattern is always the same. Every odd number is followed by an even number.

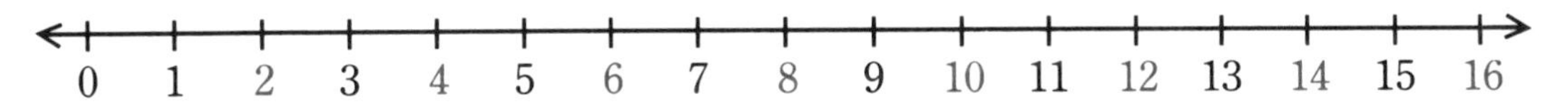

PRACTICE

Every other number is an odd number.

A. Look at the numbers below. Read them from left to right. The first odd number is 1. Copy only the odd numbers on a separate sheet of paper.

1	2	3	4	5	6	7	8	9	10
11	12	13	14	15	16	17	18	19	20
21	22	23	24	25	26	27	28	29	30
31	32	33	34	35	36	37	38	39	40
41	42	43	44	45	46	47	48	49	50
51	52	53	54	55	56	57	58	59	60

B. Look at the numbers below. Read them from left to right. The first even number is 52. Copy only the even numbers on a separate sheet of paper.

51	52	53	54	55	56	57	58	59	60
61	62	63	64	65	66	67	68	69	70
71	72	73	74	75	76	77	78	79	80
81	82	83	84	85	86	87	88	89	90
91	92	93	94	95	96	97	98	99	100
101	102	103	104	105	106	107	108	109	110

1.4 Place Value and Renaming

The ten **digits** are 0, 1, 2, 3, 4, 5, 6, 7, 8, and 9. You can use these digits to write any number. The important thing is to put each digit in the correct place.

Tens and Ones

Look where the digits in the number 24 are placed.

To find out what a digit is worth, notice where it is placed.

The 2 in 24 means two tens. It is in the tens place.
The 4 means four ones. It is in the ones place.

The number 24 can be **renamed** two tens and four ones or two tens + four ones.

Now look at a different number with the same digits: 42.

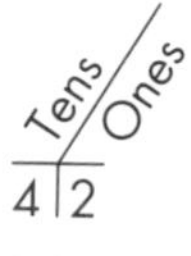

In this number the 4 means four tens. It is in the tens place. The 2 means two ones. It is in the ones place.

The number 42 can be *renamed* four tens + two ones.

PRACTICE

Copy the number sentences on a separate sheet of paper. Put the correct number words in the blanks. The first one has been done for you.

1. 23 can be renamed *two* tens and *three* ones.
2. 17 can be renamed ________ tens + ________ ones.
3. 69 can be renamed ________ tens + ________ ones.
4. 46 can be renamed ________ tens + ________ ones.
5. 81 can be renamed ________ tens + ________ ones.
6. 55 can be renamed ________ tens + ________ ones.

7. 70 can be renamed ________ tens + ________ ones.

8. 98 can be renamed ________ tens + ________ ones.

9. 36 can be renamed ________ tens + ________ ones.

10. 57 can be renamed ________ tens + ________ ones.

Thousands and Hundreds

Look at the digits in the number 3,724. Notice where each digit is placed.

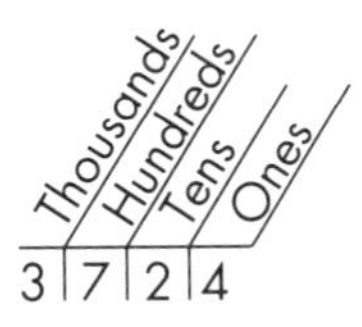

The 3 means three thousands. The 3 is in the thousands place.
The 7 means seven hundreds. The 7 is in the hundreds place.
The 2 means two tens. The 2 is in the tens place.
The 4 means four ones. The 4 is in the ones place.

The number 3,724 can be *renamed* this way: three thousands + seven hundreds + two tens + four ones.

PRACTICE

A. Copy the number sentences on a separate sheet of paper. Put the correct number words in the blanks. The first one has been done for you.

1. 5,200 can be renamed *five* thousands + *two* hundreds.

2. 1,800 can be renamed ________ thousand(s) + ________ hundreds.

3. 6,600 can be renamed ________ thousands + ________ hundreds.

4. 3,100 can be renamed ________ thousands + ________ hundred(s).

5. 9,700 can be renamed ________ thousands + ________ hundreds.

B. Copy the number sentences on a separate sheet of paper. Put the correct number words in the blanks.

1. 2,423 means ________ thousands + ________ hundreds + ________ tens + ________ ones.
2. 7,502 means ________ thousands + ________ hundreds + ________ tens + ________ ones.
3. 5,349 means ________ thousands + ________ hundreds + ________ tens + ________ ones.
4. 8,137 means ________ thousands + ________ hundreds + ________ tens + ________ ones.
5. 6,054 means ________ thousands + ________ hundreds + ________ tens + ________ ones.

C. Read the place name before each number. On a separate sheet of paper, write only the digit that appears in that place.

1.	Ones	3,582	**6.**	Hundreds	5,210
2.	Thousands	9,043	**7.**	Tens	6,821
3.	Hundreds	2,849	**8.**	Thousands	8,103
4.	Tens	4,613	**9.**	Hundreds	7,955
5.	Ones	8,851	**10.**	Thousands	1,922

1.5 Reading and Writing Whole Numbers

Look at the chart on the following page. Notice the names of each place.

What a number is worth is called its value.

Numbers are always read from left to right. Look at the first number on the chart, 529,682.

Step 1: Look at the group of numbers in color on the chart. They are the thousands numbers. Read this group.
5 hundred 29 thousand

Step 2: Look at the group of numbers in black. Read each digit and then its place name.
6 hundred eighty-two

Step 3: Read the rest of the numbers.

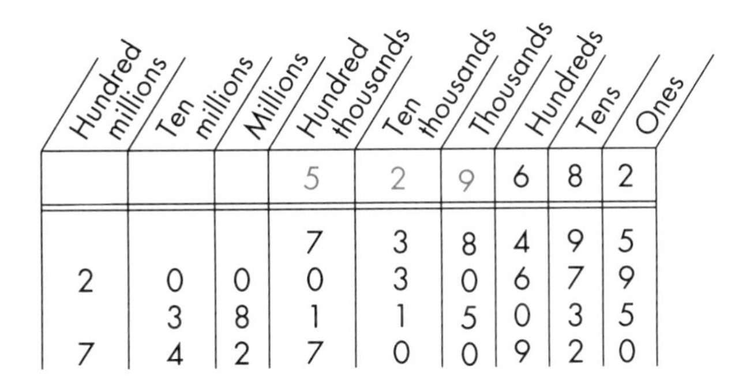

Hundred millions	Ten millions	Millions	Hundred thousands	Ten thousands	Thousands	Hundreds	Tens	Ones
			5	2	9	6	8	2
			7	3	8	4	9	5
2	0	0	0	3	0	6	7	9
	3	8	1	1	5	0	3	5
7	4	2	7	0	0	9	2	0

PRACTICE

A. Copy the sentences on a separate sheet of paper. Fill in the blanks with the correct number words.

1. 32,584 means ________ thousands + ________ hundreds + ________ tens + ________ ones.

2. 74,800,000 means ________ millions + ________ hundred thousands.

3. 6,002,070 means ________ millions + ________ thousands + ________ tens.

B. Use digits to write the numbers below on a separate sheet of paper. Put a comma after the millions digit. Put a comma after the thousands digits.

A comma looks like this: , Commas are used to set off groups of numbers just as they are used to set off groups of words.

1. three million, two hundred thousand, sixty-nine.

2. nine hundred fifty-eight thousand, six hundred.

3. one hundred fifty-two million, five hundred, two.

4. sixteen million, four hundred thousand.

5. seven hundred million, twelve.

6. five hundred sixteen thousand, ninety-five.

1.6 Rounding Whole Numbers

Rounding to the Nearest Ten

Rounding a number makes it easier to work with. When you change a number to the nearest tens number, you get a simpler number. Changing numbers to the nearest ten is called *rounding numbers.*

Look at the chart. Notice the numbers in color. These are the tens numbers.

1	2	3	4	5	6	7	8	9	10
11	12	13	14	15	16	17	18	19	20
21	22	23	24	25	26	27	28	29	30
31	32	33	34	35	36	37	38	39	40
41	42	43	44	45	46	47	48	49	50
51	52	53	54	55	56	57	58	59	60
61	62	63	64	65	66	67	68	69	70
71	72	73	74	75	76	77	78	79	80
81	82	83	84	85	86	87	88	89	90
91	92	93	94	95	96	97	98	99	100

Round this number to the nearest ten: 23. Follow the steps.

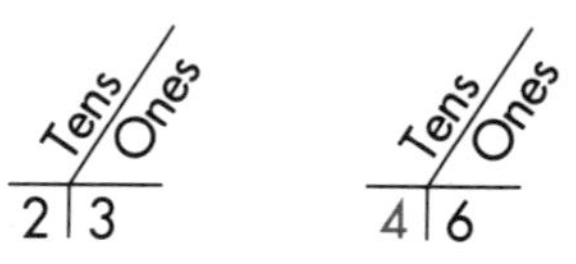

Step 1: First look at the digit in the ones place.
Step 2: Ask yourself if this digit is 5 or more.
If it is, add 1 to the tens place.
If it isn't, make no change in the tens place.
Step 3: Put a 0 in the ones place.

2 0   5 0

Round this number to the nearest ten: 46. Follow the same three steps above.

PRACTICE

A. Write the short number lines below on a separate sheet of paper. Count by tens to fill in the blanks. The first one has been done for you.

1.	10	20	30	**5.**	__	70	__	**9.**	__	17	__
2.	__	60	__	**6.**	__	69	__	**10.**	__	50	__
3.	__	34	__	**7.**	__	40	__	**11.**	__	85	__
4.	__	30	__	**8.**	__	80	__	**12.**	__	92	__

B. On a separate sheet of paper round the number in color from each group. Choose the correct answer from the three answers given.

1.	12	20	10	9	**4.**	7	17	0	10	**7.**	74	80	85	70
2.	36	40	5	30	**5.**	65	60	40	70	**8.**	4	20	0	10
3.	86	80	90	87	**6.**	59	50	40	60					

C. Round the following numbers to the nearest ten. Write the answers on a separate sheet of paper.

When you round to the nearest ten, work only with the tens and ones places.

1.	22	**8.**	88,881	**15.**	23,314
2.	379	**9.**	334	**16.**	564
3.	175	**10.**	1,661	**17.**	830,019
4.	6	**11.**	327	**18.**	53,426
5.	843	**12.**	105,302	**19.**	89
6.	2,458	**13.**	39	**20.**	4,228
7.	32,072	**14.**	387,611		

Rounding to the Nearest Hundred

		Hundreds / Tens / Ones	Thousands / Hundreds / Tens / Ones
Step 1:	First look at the tens place.	3 | 6 | 2	2 ,8 | 3 | 1
Step 2:	Ask yourself if the digit in the tens place is five or more. If it is, add 1 to the hundreds place. If it isn't, make no change.	3 6 2 +1	2 ,8 3 1
Step 3:	Change the digits to the right to 0.	4 0 0	2 ,8 0 0

Rounding to the Nearest Thousand

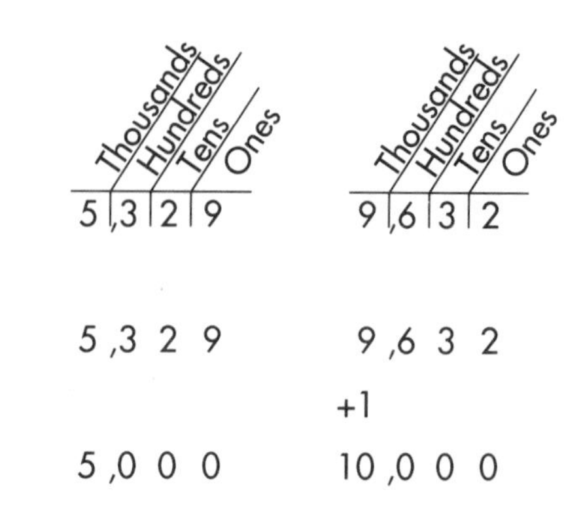

Step 1: Look at the hundreds place.

Step 2: Ask yourself if the digit in the hundreds place is five or more.
If it is, add 1 to the thousands place.
If it isn't, make no changes in the thousands place.

Step 3: Change all the digits to the right to 0.

PRACTICE

A. On a separate sheet of paper round each number in color to the nearest 100. Choose the correct answer from the three answers given.

1.	724	600	700	800
2.	539	540	600	500
3.	874	870	900	800
4.	659	600	650	700
5.	8,239	8,200	8,300	8,000
6.	7,672	7,700	7,600	8,000

B. On a separate sheet of paper round each number in color to the nearest 1000. Choose the correct answer from the three answers given.

1.	4,341	5,000	4,000	4,300
2.	8,715	7,000	8,000	9,000
3.	26,352	27,000	37,000	26,000
4.	145,479	150,000	145,000	146,000
5.	46,863	46,000	45,000	47,000
6.	9,333	9,000	10,000	9,500
7.	18,099	10,000	18,000	20,000
8.	246,588	245,000	246,000	247,000

C. Round the numbers to the nearest hundred. Write your answers on a separate sheet of paper.

1. 446	**5.** 852	**9.** 291
2. 902	**6.** 3,720	**10.** 5,555
3. 4,670	**7.** 993	**11.** 3,082
4. 1,963	**8.** 607	**12.** 930

D. Round the numbers to the nearest thousand. Write your answers on a separate sheet of paper.

1. 3,522	**5.** 7,298	**9.** 6,700
2. 8,050	**6.** 9,825	**10.** 1,298
3. 33,694	**7.** 29,547	**11.** 88,027
4. 10,713	**8.** 6,200	**12.** 50,514

E. Number your paper from 1 to 10. Copy the numbers below. Then round each one three times. First, round the number to the nearest ten. Next, round the number to the nearest hundred. Then round the number to the nearest thousand. The first one has been done for you.

Notice how many 0's there are each time you round. How many are there when you round up to tens? To hundreds? To thousands?

1. 3,256 3,260 3,300 3,000
2. 4,641
3. 9,897
4. 12,560
5. 29,705
6. 43,489
7. 72,555
8. 68,072
9. 89,394
10. 53,627

F. To which place is each number rounded? The first one has been done for you.

1. 15,863 16,000 thousands
2. 11,253 11,300
3. 4,757 4,760
4. 75,362 75,000
5. 8,955 9,000

MATHEMATICS IN YOUR LIFE:
Numbers on Labels

"My doctor says the only thing that is wrong with me is the way I eat," said Marge. "She wants me to eat fewer calories, less fat, and less sodium. But how am I to know how many calories are in packaged foods?"

"You can find out what you need to know by reading the labels," answered Ginny. "This label says this can of stewed tomatoes contains 4 servings. Each serving has only 35 calories and no fat. That's good. But it does have quite a bit of sodium—360 milligrams. Maybe you'd better look for a can of low-sodium tomatoes."

The labels below give some of the information found on packaged foods. Read the labels. Then answer the questions.

a.

SERVING SIZE .... 1 CUP
SERVINGS 4
CALORIES PER SERVING 35
FAT 0
SODIUM .... 360 MILLIGRAMS

b.

SERVING SIZE .... 1 CUP
SERVINGS 4
CALORIES PER SERVING .. 240
FAT .... 22 GRAMS
SODIUM .... 160 MILLIGRAMS

c.

SERVING SIZE .... 1 CUP
SERVINGS 6
CALORIES PER SERVING 28
FAT 0
SODIUM .... 420 MILLIGRAMS

d.

SERVING SIZE .... 1 CUP
SERVINGS 4
CALORIES PER SERVING 40
FAT .... 10 GRAMS
SODIUM ...... 75 MILLIGRAMS

1. Which label shows the fewest calories?
2. Which label shows the most calories?
3. Which labels show the least fat?
4. Which label shows the most fat?
5. Which label shows the least sodium?
6. Which label shows the most sodium?

1.7 How Your Calculator Works

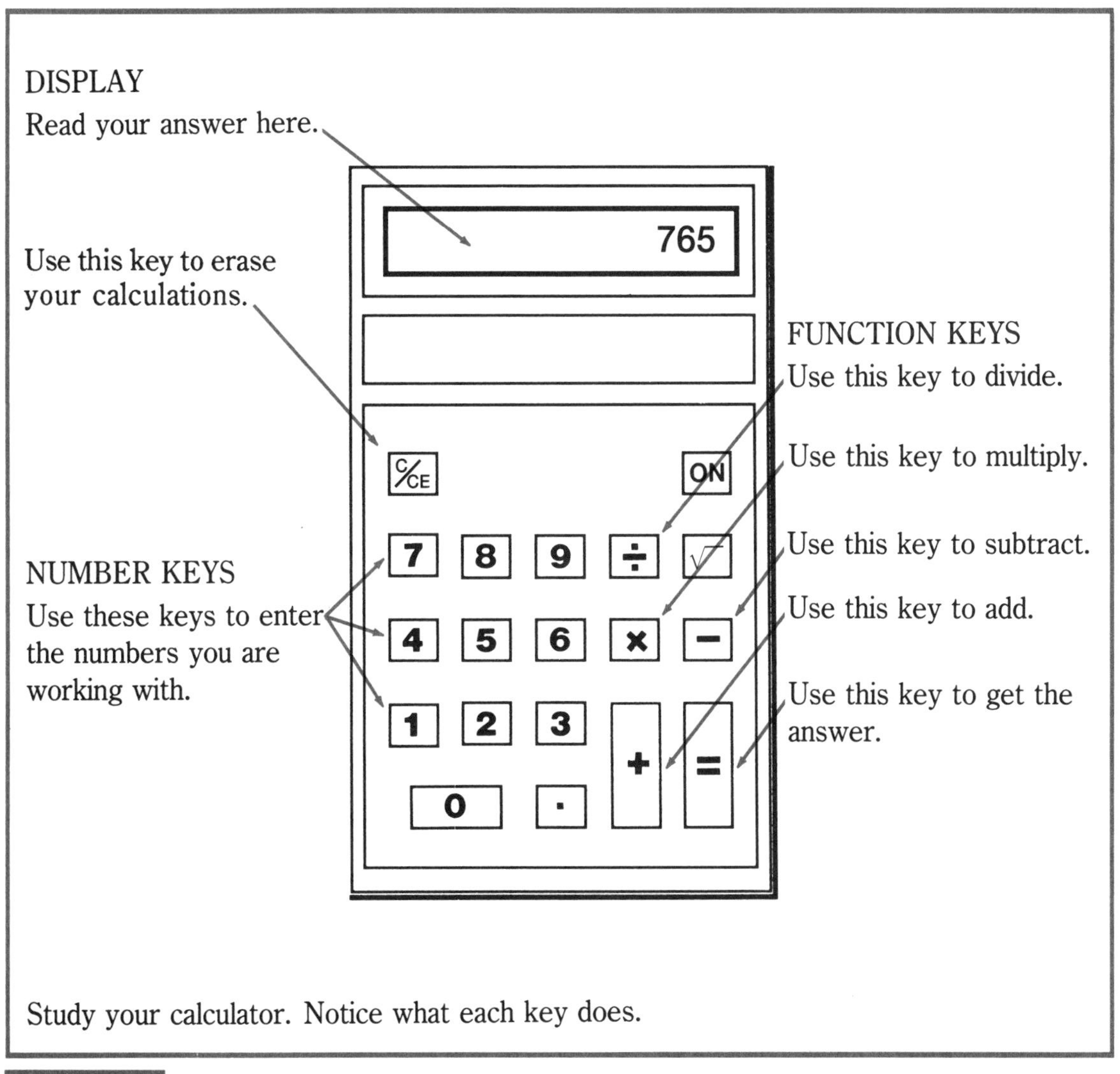

Study your calculator. Notice what each key does.

PRACTICE

Enter each number on your calculator. Then erase it.

1.	345	**6.**	87,642
2.	973	**7.**	341,829
3.	9,275	**8.**	757,691
4.	8,787	**9.**	2,590,327
5.	27,385	**10.**	9,542,611

CHAPTER REVIEW

CHAPTER SUMMARY

■ Whole Numbers	The numbers 0, 1, 2, 3, and so on are whole numbers.
■ Number Line	A number line shows the numbers in order. The numbers get bigger from left to right.
■ Odd and Even Numbers	The numbers 1, 3, 5, 7, and so on are odd numbers. The numbers 2, 4, 6, 8, and so on are even numbers. Every odd number is followed by an even number.
■ Place Value	You can write any number with the ten digits: 0, 1, 2, 3, 4, 5, 6, 7, 8, 9. The number 4 in the ones place means four ones. The number 4 in the tens place means four tens. The number 4 in the hundreds place means four hundreds. The number 4 in the thousands place means four thousands.
■ Rounding Numbers	Rounding numbers to the nearest ten, hundred, or thousand makes numbers easier to work with.

REVIEWING VOCABULARY

Copy the sentences on a separate sheet of paper. Put in the correct word or group of words from the box.

whole number	rounded	digits	rename	even	odd

1. 234 is a ________ ________ .

2. 0, 1, 2, 3, and 4 are ________ .

3. You can ________ 52 as five tens + two ones.

4. 37 ________ to the nearest ten is 40.

5. The numbers 2, 4, 6 and 8 are ________ numbers.

6. The numbers 1, 3, 5 and 7 are ________ numbers.

CHAPTER REVIEW

CHAPTER QUIZ

A. Copy the number line below on a separate sheet of paper.

1. Fill in each blank with the correct number.

30 __ 32 33 34 __ 36 37 __ 39 __ 41 __

2. Look at the number line you just completed. List only the odd numbers.

B. Rename these numbers on a separate sheet of paper.

1. 23 means ________ tens + ________ ones.

2. 56 means ________ tens + ________ ones.

3. 198 means ________ hundreds + ________ tens + ________ ones.

4. 402 means ________ hundreds + ________ tens + ________ ones.

5. 1,111 means ________ thousands + ________ hundreds + ________ tens + ________ ones.

C. On a separate sheet of paper, round each number to the nearest ten.

1. 653
2. 207
3. 88
4. 6,338
5. 7,888
6. 250

D. On a separate sheet of paper, round each number to the nearest hundred.

1. 292
2. 763
3. 992
4. 4,058
5. 7,213
6. 9,641

E. On a separate sheet of paper, round each number to the nearest thousand.

1. 8,566
2. 38,099
3. 988
4. 11,256
5. 87,497
6. 7,777

ADDING WHOLE NUMBERS

2

These are just a few of the thousands of shelves in the central public library of Los Angeles. The library houses more than 2 million books. Each year approximately 50,000 new books are added to its stacks.

Chapter Learning Objectives

1. Add whole numbers
2. Add zero
3. Add larger numbers
4. Rename numbers
5. Estimate answers
6. Use addition to solve word problems

WORDS TO KNOW

- **add** to put numbers together; to find the total amount
- **equal** the same as; =
- **sum** the amount obtained by adding; the total
- **estimate** to quickly figure out an answer that is close to the exact answer; to make a good guess
- **column** numbers placed one below the other
- **compare** to see how two things are alike or different
- **solve** to find the answer to a problem

2.1 What Is Addition?

Addition is the process of putting numbers together to get a total. The answer you get in an addition problem is called the **sum.** The drawing below shows three ways to **add:** in pictures, in words, and in numbers.

A process is the set of steps it takes to do something.

■■		■■■		■■■■■
two	plus	three	**equals**	five
2	+	3	=	5

In this chapter you will learn more about addition. You will also learn when and how to use addition to solve word problems.

PRACTICE

Write these addition problems using numbers. The first one has been done for you.

1. Eight plus three equals eleven $8 + 3 = 11$
2. Four plus five equals nine
3. Six plus seven equals thirteen
4. Two plus ten equals twelve

2.2 Basic Addition

Reading and Writing Addition Examples

Addition problems may be written two ways. Read the examples below.

3	+	4	=	7
Three	plus	four	equals	seven.

$$\begin{array}{rl} 3 & \text{Three} \\ +4 & \text{plus four} \\ \hline & \text{equals} \\ 7 & \text{seven.} \end{array}$$

Order in Addition

You can add whole numbers in any order. The sum will not change.

Example: $2 + 3 = 5$ $3 + 2 = 5$
$5 + 4 = 9$ $4 + 5 = 9$

PRACTICE

Read each example.

1. $4 + 2 = 6$

2. $2 + 4 = 6$

3. $7 + 8 + 2 = 17$

4. $6 + 5 + 4 + 2 = 17$

5. $\begin{array}{r} 9 \\ +5 \\ \hline 14 \end{array}$

6. $\begin{array}{r} 7 \\ +7 \\ \hline 14 \end{array}$

7. $\begin{array}{r} 3 \\ +5 \\ \hline 8 \end{array}$

8. $\begin{array}{r} 6 \\ +6 \\ \hline 12 \end{array}$

9. $8 + 3 + 4 = 15$

10. $4 + 5 + 9 = 18$

11. $5 + 6 + 2 + 0 = 13$

12. $1 + 6 + 2 + 5 = 14$

Using a Number Line to Add

Numbers can be shown as points on a line. A number line can help you add.

Look at the example and the number line.

Example:
5 + 3 = 8

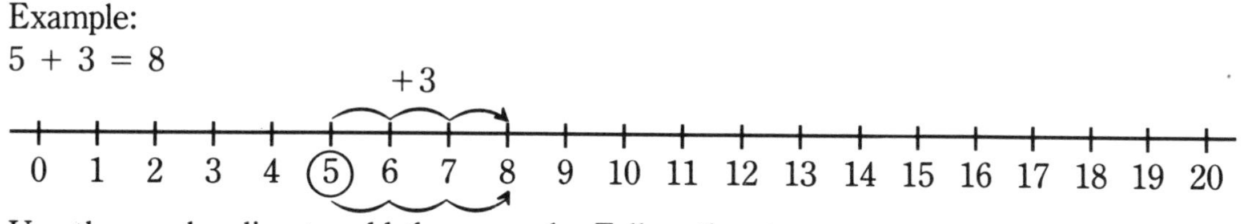

Use the number line to add the example. Follow the steps below.

Step 1: 5 is the first number in the example. Find 5 on the number line.

Step 2: You need to add 3. Move three numbers to the right. You are at 8. The number line shows that 5 + 3 = 8.

PRACTICE

A. Copy the problems below on a separate sheet of paper. Use the number line to help you find the sums.

1. 9 + 1 =
2. 6 + 7 =
3. 4 + 2 =
4. 5 + 7 =
5. 4 + 9 =
6. 8 + 9 =
7. 8 + 7 =
8. 3 + 2 =
9. 3 + 6 =
10. 8 + 6 =
11. 7 + 2 =
12. 9 + 7 =
13. 3 + 8 =
14. 7 + 5 =
15. 5 + 5 =
16. 5 + 8 =
17. 2 + 9 =
18. 4 + 8 =
19. 7 + 8 =
20. 9 + 4 =
21. 5 + 9 =
22. 7 + 3 =
23. 7 + 7 =
24. 4 + 6 =
25. 8 + 8 =
26. 5 + 6 =
27. 9 + 9 =
28. 9 + 8 =
29. 2 + 7 =
30. 3 + 5 =
31. 4 + 4 =
32. 1 + 9 =
33. 3 + 3 =
34. 8 + 5 =
35. 7 + 4 =
36. 8 + 4 =
37. 6 + 4 =
38. 9 + 5 =
39. 6 + 6 =
40. 2 + 5 =

B. Copy the problems below on a separate sheet of paper. Use the number line to help you find the sums.

1. $\begin{array}{r} 3 \\ +1 \\ \hline \end{array}$

2. $\begin{array}{r} 5 \\ +4 \\ \hline \end{array}$

3. $\begin{array}{r} 6 \\ +9 \\ \hline \end{array}$

4. $\begin{array}{r} 6 \\ +5 \\ \hline \end{array}$

5. $\begin{array}{r} 7 \\ +4 \\ \hline \end{array}$

6. $\begin{array}{r} 4 \\ +3 \\ \hline \end{array}$

7. $\begin{array}{r} 9 \\ +9 \\ \hline \end{array}$

8. $\begin{array}{r} 8 \\ +6 \\ \hline \end{array}$

9. $\begin{array}{r} 5 \\ +3 \\ \hline \end{array}$

10. $\begin{array}{r} 9 \\ +7 \\ \hline \end{array}$

11. $\begin{array}{r} 7 \\ +6 \\ \hline \end{array}$

12. $\begin{array}{r} 8 \\ +8 \\ \hline \end{array}$

13. $\begin{array}{r} 2 \\ +7 \\ \hline \end{array}$

14. $\begin{array}{r} 6 \\ +6 \\ \hline \end{array}$

15. $\begin{array}{r} 5 \\ +4 \\ \hline \end{array}$

16. $\begin{array}{r} 9 \\ +5 \\ \hline \end{array}$

17. $\begin{array}{r} 8 \\ +7 \\ \hline \end{array}$

18. $\begin{array}{r} 3 \\ +9 \\ \hline \end{array}$

19. $\begin{array}{r} 4 \\ +8 \\ \hline \end{array}$

20. $\begin{array}{r} 7 \\ +9 \\ \hline \end{array}$

Adding Zero

When you add zero to any whole number, the number stays the same.

Read these examples.

$6 + 0 = 6$ $\quad 0 + 6 = 6$ $\quad 9 + 0 = 9$ $\quad 0 + 9 = 9$

PRACTICE

Copy the problems below on a separate sheet of paper. Find the sums. If you need help, use the number line on the previous page.

1. $3 + 0 =$

2. $5 + 4 =$

3. $1 + 7 =$

4. $6 + 7 =$

5. $6 + 4 =$

6. $7 + 9 =$

7. $8 + 0 =$

8. $7 + 8 =$

9. $0 + 7 =$

10. $0 + 0 =$

11. $5 + 7 =$

12. $9 + 4 =$

13. $3 + 9 =$

14. $8 + 4 =$

15. $4 + 8 =$

16. $0 + 8 =$

17. $8 + 3 =$

18. $2 + 9 =$

19. $0 + 5 =$

20. $7 + 8 =$

2.3 Adding a Single Column

Add the numbers in each example. Follow the steps below.

Step 1: Add 3 + 4 = 7
Step 2: Add 7 + 2 = 9

$$\begin{array}{r} 3 \\ 4 \\ +2 \\ \hline 9 \end{array}$$

Step 1: Add 2 + 3 = 5
Step 2: Add 5 + 4 = 9
Step 3: Add 9 + 6 = 15

$$\begin{array}{r} 2 \\ 3 \\ 4 \\ +6 \\ \hline 15 \end{array}$$

PRACTICE

A. Copy these problems on a separate sheet of paper. Find the sums.

1. $\begin{array}{r} 3 \\ 1 \\ +2 \\ \hline \end{array}$ **3.** $\begin{array}{r} 6 \\ 4 \\ +9 \\ \hline \end{array}$ **5.** $\begin{array}{r} 5 \\ 3 \\ +7 \\ \hline \end{array}$ **7.** $\begin{array}{r} 2 \\ 7 \\ +6 \\ \hline \end{array}$ **9.** $\begin{array}{r} 4 \\ 9 \\ +3 \\ \hline \end{array}$

2. $\begin{array}{r} 2 \\ 5 \\ 4 \\ +8 \\ \hline \end{array}$ **4.** $\begin{array}{r} 8 \\ 2 \\ 6 \\ +3 \\ \hline \end{array}$ **6.** $\begin{array}{r} 4 \\ 7 \\ 3 \\ +2 \\ \hline \end{array}$ **8.** $\begin{array}{r} 5 \\ 6 \\ 3 \\ +6 \\ \hline \end{array}$ **10.** $\begin{array}{r} 7 \\ 9 \\ 1 \\ +0 \\ \hline \end{array}$

B. On a separate sheet of paper, write each group of numbers in columns. Then find the sums.

1. 3, 9, 2
2. 5, 1, 2, 8
3. 7, 3, 4, 5
4. 5, 3, 7
5. 4, 5, 1, 8
6. 3, 9, 6, 2
7. 6, 9, 2
8. 9, 3, 5, 3
9. 5, 1, 6, 4
10. 6, 2, 5, 3
11. 7, 8, 7, 5
12. 2, 3, 6, 2
13. 4, 8, 2, 5
14. 6, 3, 7, 8
15. 5, 9, 3, 4
16. 6, 9, 0, 5
17. 4, 3, 8, 0
18. 5, 0, 5, 8

2.4 Adding Larger Numbers

Reviewing Place Value

Study how the numbers are written.
4 and 5 are in the ones place.
0 and 7 are in the tens place.
3 and 6 are in the hundreds place.

$$\begin{array}{r} 304 \\ +675 \\ \hline \end{array}$$

Find the sum. $\begin{array}{r} 304 \\ +675 \\ \hline \end{array}$ Follow the steps below.

Step 1: Add the digits in the ones place.
Step 2: Add the digits in the tens place.
Step 3: Add the digits in the hundreds place.

Hundreds	Tens	Ones
3	0	4
+6	7	5
9	7	9

PRACTICE

A. Write the problems below on a separate sheet of paper. Find the sums.

1. $\begin{array}{r} 30 \\ +10 \\ \hline \end{array}$ **4.** $\begin{array}{r} 20 \\ +45 \\ \hline \end{array}$ **7.** $\begin{array}{r} 50 \\ +30 \\ \hline \end{array}$ **10.** $\begin{array}{r} 90 \\ +\ 7 \\ \hline \end{array}$ **13.** $\begin{array}{r} 35 \\ +40 \\ \hline \end{array}$

2. $\begin{array}{r} 37 \\ +41 \\ \hline \end{array}$ **5.** $\begin{array}{r} 95 \\ +\ 3 \\ \hline \end{array}$ **8.** $\begin{array}{r} 47 \\ +42 \\ \hline \end{array}$ **11.** $\begin{array}{r} 64 \\ +23 \\ \hline \end{array}$ **14.** $\begin{array}{r} 73 \\ +15 \\ \hline \end{array}$

3. $\begin{array}{r} 41 \\ 27 \\ +11 \\ \hline \end{array}$ **6.** $\begin{array}{r} 35 \\ 42 \\ +12 \\ \hline \end{array}$ **9.** $\begin{array}{r} 26 \\ 10 \\ +51 \\ \hline \end{array}$ **12.** $\begin{array}{r} 13 \\ 43 \\ +\ 2 \\ \hline \end{array}$ **15.** $\begin{array}{r} 27 \\ 22 \\ +40 \\ \hline \end{array}$

B. Write the problems below on a separate sheet of paper. Find the sums.

1. $\begin{array}{r} 200 \\ +705 \\ \hline \end{array}$ **2.** $\begin{array}{r} 628 \\ +251 \\ \hline \end{array}$ **3.** $\begin{array}{r} 3,562 \\ +5,420 \\ \hline \end{array}$ **4.** $\begin{array}{r} 2,318 \\ +3,641 \\ \hline \end{array}$ **5.** $\begin{array}{r} 3,326 \\ +2,213 \\ \hline \end{array}$

2.5 Adding with One Renaming

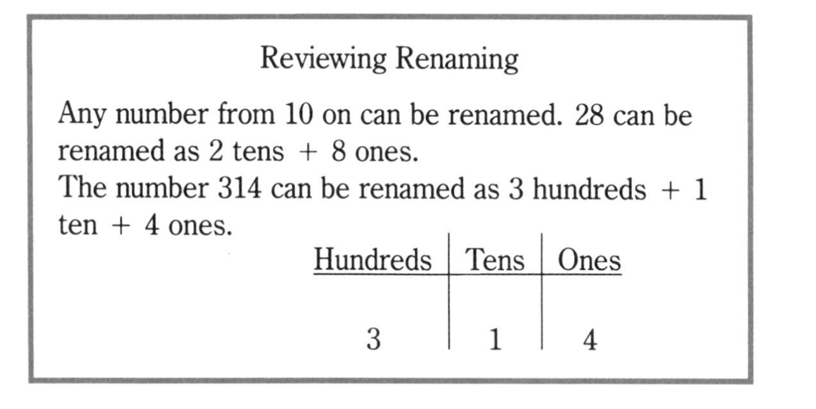

Reviewing Renaming

Any number from 10 on can be renamed. 28 can be renamed as 2 tens + 8 ones.
The number 314 can be renamed as 3 hundreds + 1 ten + 4 ones.

Hundreds	Tens	Ones
3	1	4

Find the answer to this problem: $\begin{array}{r} 48 \\ +24 \\ \hline \end{array}$
Follow the steps below.

Step 1: Add the digits in the ones place. 8 + 4 = 12
Step 2: Rename 12. 1 ten + 2 ones.
Step 3: Write 2 in the ones place.
Write 1 above the tens place.
Step 4: Add the digits in the tens place. 1 + 4 + 2 = 7
Write 7 in the tens place.

Tens	Ones
(1)	
4	8
+2	4
7	2

PRACTICE

A. Copy the problems below on a separate sheet of paper. Find the sums. Don't forget to rename.

1. $\begin{array}{r} 28 \\ +\ 3 \\ \hline \end{array}$ **4.** $\begin{array}{r} 35 \\ +57 \\ \hline \end{array}$ **7.** $\begin{array}{r} 67 \\ +23 \\ \hline \end{array}$ **10.** $\begin{array}{r} 48 \\ +23 \\ \hline \end{array}$ **13.** $\begin{array}{r} 83 \\ +19 \\ \hline \end{array}$

2. $\begin{array}{r} 25 \\ +46 \\ \hline \end{array}$ **5.** $\begin{array}{r} 73 \\ +\ 9 \\ \hline \end{array}$ **8.** $\begin{array}{r} 59 \\ +24 \\ \hline \end{array}$ **11.** $\begin{array}{r} 48 \\ +\ 9 \\ \hline \end{array}$ **14.** $\begin{array}{r} 46 \\ +48 \\ \hline \end{array}$

3. $\begin{array}{r} 49 \\ +45 \\ \hline \end{array}$ **6.** $\begin{array}{r} 19 \\ +72 \\ \hline \end{array}$ **9.** $\begin{array}{r} 86 \\ +\ 5 \\ \hline \end{array}$ **12.** $\begin{array}{r} 67 \\ +35 \\ \hline \end{array}$ **15.** $\begin{array}{r} 66 \\ +29 \\ \hline \end{array}$

B. Copy the problems below on a separate sheet of paper. Find the sums. You will not need to rename in all the problems.

1. 16 + 7

2. 37 + 55

3. 48 + 47

4. 84 + 8

5. 27 + 72

6. 49 + 39

7. 32 + 59

8. 66 + 23

9. 54 + 25

10. 57 + 36

2.6 Adding with More Than One Renaming

The example below has more than one renaming.

 584
+977

Follow the steps and find the sum.

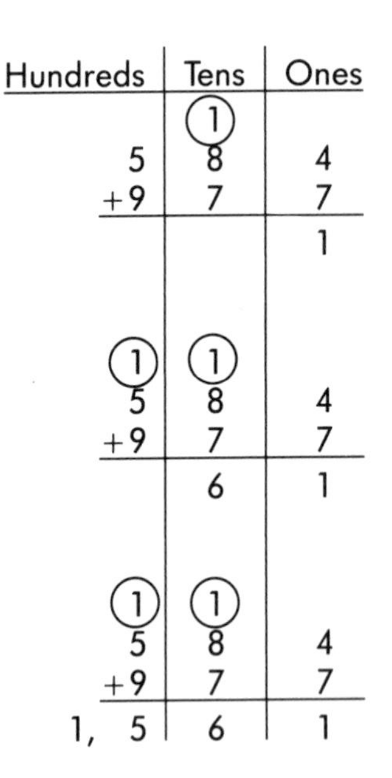

Step 1: Add the numbers in the ones place.
$4 + 7 = 11$

Step 2: Rename 11. 1 ten + 1 one

Step 3: Write 1 in the ones place.
Write 1 above the tens place.

Step 4: Add the numbers in the tens place.
$1 + 8 + 7 = 16$

Step 5: Rename 16 (in the tens place) 1 hundred + 6 tens.

Step 6: Write 6 in the tens place.

Step 7: Add the numbers in the hundreds place.
$1 + 5 + 9 = 15$

Step 8: Rename 15 (in the hundreds place)
1 thousand + 5 hundreds

Step 9: Write 5 in the hundreds place.
Write 1 in the thousands place.

PRACTICE

A. Copy the problems below on a separate sheet of paper. Find the sums. Don't forget to rename.

1. 256 + 375	**3.** 489 + 314	**5.** 687 + 124	**7.** 565 + 88	**9.** 337 + 596
2. 798 + 388	**4.** 546 + 773	**6.** 769 + 874	**8.** 547 + 495	**10.** 598 + 512

B. Copy the problems below on a separate sheet of paper. Find the sums. Don't forget to rename.

1. 287 + 354	**2.** 564 + 769	**3.** 139 + 825	**4.** 805 + 562	**5.** 693 + 174

2.7 Estimating and Thinking

Estimating will not give you an exact answer to a problem, but it will quickly give you an answer that is close to the exact one. Use round numbers to estimate.

Reviewing Rounding Numbers

	Rounding to the Nearest Ten	Hundred	Thousand
Choose the place you are rounding to.	34	457	1,852
Is the digit to the right 5 or more?	34	457	1,852
If it is, add to the place you are rounding to.		+1	+1
If it isn't, make no changes.			
Replace the digits to the right with zeros.	30	500	2,000

Follow the steps below to help you estimate the answer to the example.

Step 1: Round the numbers to the nearest ten.
Step 2: Add the rounded numbers.
Step 3: Add the exact numbers.
Step 4: **Compare** the sums.

Exact	Rounded
48	50
31	30
79	80

PRACTICE

A. Copy the problems below on a separate sheet of paper. Round the numbers to the nearest ten. Estimate the answers. Then find the exact sums.

1. 41 + 23 **2.** 36 + 32 **3.** 65 + 17 **4.** 48 + 26

B. Copy the problems below on a separate sheet of paper. Round the numbers to the nearest hundred. Estimate the answers. Then find the exact sums.

1. 126 + 452 **2.** 588 + 234 **3.** 291 + 763 **4.** 374 + 588 **5.** 835 + 656

2.8 Solving Addition Problems

Solving word problems is much easier if you do each step in order.

Steps in Solving Word Problems

Read this problem. Follow the steps below to solve it.

Jerry drove 328 miles on Monday. He drove 207 miles on Tuesday. On Wednesday, he drove 193 miles. How many miles did he drive in all?

Step 1: Read the problem.
Step 2: Learn what you must find out.
Step 3: Notice the clue words. Clue words give you the key to solving the problem. *In all* tells you to add.
Step 4: Write down the numbers.
Step 5: Add the columns.
Step 6: Check the answer. Ask yourself if it makes sense.

How many miles did Jerry drive in all?
Clue words: in all

Numbers: 328 207 193

328
207
+193
728

Clue Words

Words and groups of words like these give you a clue. They tell you to put numbers together. They tell you to add.

in all *together* *altogether* *total* *both*

Reminder: Not all problems have clue words.

PROBLEMS TO SOLVE

Solve these problems on a separate sheet of paper.

All of these problems can be solved by using the steps on page 28.

1. Nick made 15 sales in the morning. He made 8 in the afternoon. How many sales did he make in all?
2. Dan typed 9 letters. Maria typed 12. How many letters did they both type?
3. The first bus carried 47 students. The second bus carried 62 students. The third bus carried 59 students. How many students were carried on all three buses?
4. Terry made 24 tuna sandwiches for the party. She made 15 cheese sandwiches. She also made 12 chicken sandwiches. How many sandwiches did she make?
5. Roger ran 3 miles on Tuesday. He ran 4 miles on Wednesday. On Friday, he ran 1 mile. On Saturday, he ran 5 miles. How many total miles did Roger run?
6. Margo put 352 books away. Later, she put 49 more books away. How many books did she put away?
7. Pedro worked 5 hours on Monday. He worked 6 hours on Wednesday. On Saturday he worked 8 hours. How many hours did he work in all?
8. Many people came to the job fair. In the morning 689 people came. In the afternoon 427 came. How many people came in all?

MATHEMATICS IN YOUR LIFE:
Counting Calories

Paula wants to lose weight. She counts the calories of the food she eats at each meal. Here is what she ate on Wednesday.

Breakfast	Calories	Lunch	Calories	Dinner	Calories
melon	80	fish	295	chicken	295
1 egg	79	rice	100	broccoli	32
bread	108	berries	45	pear	100
coffee	0	coffee	0	coffee	0

Answer the questions below on a separate sheet of paper.

1. How many calories did Paula eat at breakfast?
2. How many calories did she eat at lunch?
3. How many did she eat at dinner?
4. At which meal did she eat the largest number of calories?
5. How many calories did she eat all day?
6. How many calories did Paula eat for breakfast and lunch?
7. How many vegetable calories did Paula eat altogether?
8. How many fruit calories did Paula eat altogether?
9. How many calories did Paula eat for lunch and dinner?
10. How many calories did Paula eat for breakfast and dinner?

2.9 Using Your Calculator: Addition

A calculator makes adding large numbers easy.
Follow the steps below to add the numbers.

Step 1: Turn on your calculator. Find the [+] sign. Find the [=] sign.

Step 2: Press [5] [8] [7] [3]

Step 3: Press [+]

Step 4: Press [3] [6] [1] [9]

Step 5: Press [=]

Step 6: Your answer should be: 9,492

PRACTICE

A. Use a calculator to find the sum of each group of numbers. Write your answers on a separate sheet of paper.

1. 3,256, 7,125
2. 7,021, 4,962, 3,332
3. 8,502, 5,843
4. 4,231, 9,500, 6,679
5. 2,910, 6,471
6. 367, 4,412

B. Use a calculator to find the sum of each column of numbers. Write your answers on a separate sheet of paper.

1.
```
   653
   295
+8,633
```

2.
```
 7,858
   263
+8,924
```

3.
```
 6,423
 5,966
+  704
```

4.
```
   875
 2,005
+  643
```

CHAPTER SUMMARY

■ Add	To add means to put numbers together; to find the total amount.	$5 + 3 = 8$
■ Adding 0	When 0 is added to a number, the number stays the same.	$5 + 0 = 5$
■ Adding Larger Numbers	To add larger numbers: Start with the digits in the ones place. Add one place at a time.	$\begin{array}{r} 345 \\ +421 \\ \hline 766 \end{array}$
■ Renaming	Sometimes digits in a place add up to 10 or more. In that case, you must rename to find the sum.	$\begin{array}{r} 257 \\ +364 \\ \hline 621 \end{array}$
■ Estimating	Estimating gives you a quick, rough idea of what an answer will be. Rounding numbers helps you estimate.	
■ Solving Addition Problems Step-by-Step	**1.** Read the problem. **2.** Learn what you must find out. **3.** Find the numbers and the clue words. **4.** Add to find the sum.	

REVIEWING VOCABULARY

Number a separate sheet of paper from 1 to 8. Read each word in the column on the left. Find its definition in the column on the right. Write the letter of the definition next to each number.

1. rename	**a.** to make a guess about the value of something or the answer to a problem
2. equal	**b.** to put numbers together; to find the total amount
3. add	**c.** to break down a number into its parts
4. compare	**d.** numbers placed one below the other
5. solve	**e.** the same as
6. column	**f.** the amount obtained by adding; the total
7. estimate	**g.** see how two things are alike or different
8. sum	**h.** to find the answer to a problem

CHAPTER REVIEW

CHAPTER QUIZ

A. Copy the problems on a separate sheet of paper. Find the sums.

1. $\begin{array}{r} 32 \\ +57 \\ \hline \end{array}$

2. $\begin{array}{r} 459 \\ +132 \\ \hline \end{array}$

3. $\begin{array}{r} 605 \\ +224 \\ \hline \end{array}$

4. $\begin{array}{r} 278 \\ +464 \\ \hline \end{array}$

5. $\begin{array}{r} 822 \\ +\ \ 43 \\ \hline \end{array}$

6. $\begin{array}{r} 796 \\ +\ \ 52 \\ \hline \end{array}$

7. $\begin{array}{r} 4 \\ 7 \\ +2 \\ \hline \end{array}$

8. $\begin{array}{r} 938 \\ +172 \\ \hline \end{array}$

9. $\begin{array}{r} 6 \\ 5 \\ +9 \\ \hline \end{array}$

10. $\begin{array}{r} 698 \\ +880 \\ \hline \end{array}$

B. Developing Thinking Skills

Read the problems. Then answer the questions on a separate sheet of paper.

Leni had 52 stamps. Chris had 69 stamps. How many did they have together?

1. What are you asked to find out?
2. What is the clue word?
3. What does the clue word tell you to do?
4. What are the numbers?
5. What is the answer to the problem?

Melissa had 46 pencils. Alex had 29 pencils. Mark had 38 pencils. How many did they have altogether?

6. What you are asked to find out?
7. What is the clue word?
8. What does the clue word tell you to do?
9. What are the numbers?
10. What is the answer to the problem?

SUBTRACTING WHOLE NUMBERS

3

This very, very rare bird is a California condor. Once thousands of these birds ranged all over North America. Over the years, their numbers have been greatly reduced. Today there are only 29 of these condors left in existence.

Chapter Learning Objectives

1. Subtract whole numbers
2. Subtract zero
3. Subtract larger numbers
4. Subtract with renaming
5. Estimate answers to subtraction problems
6. Use subtraction to solve word problems

WORDS TO KNOW

- **subtract** to take away one number from another; to find the difference between two numbers
- **difference** the amount by which one number is larger or smaller than another; the amount that remains after one number is subtracted from another

3.1 What Is Subtraction?

Subtraction is the process of taking one number away from another. The answer you get in a subtraction problem is called the **difference.**

Can you name a time when you had to subtract to find an answer?

Below are three ways of showing subtraction: in words, in pictures, and in numbers.

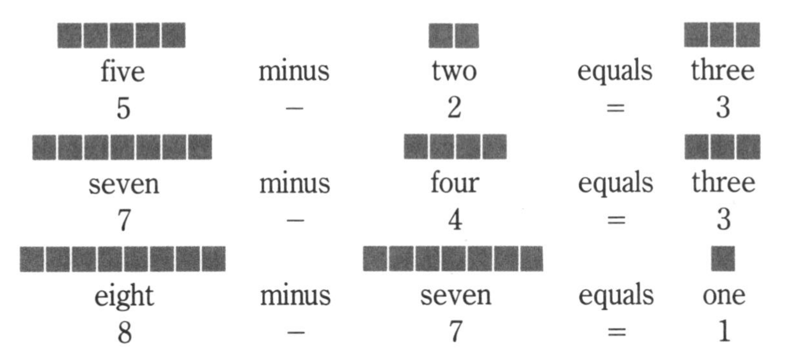

In this chapter you will learn more about subtraction. You will also learn when and how to use subtraction to solve word problems.

PRACTICE

Write these subtraction problems using numbers. The first one has been done for you.

1. Nine minus three equals six $9 - 3 = 6$
2. Eight minus five equals three
3. Five minus one equals four

3.2 Basic Subtraction

Reading and Writing Subtraction Examples

Subtraction problems may be written in two ways. Read the examples below.

7	−	4	=	3
seven	minus	four	equals	three

$$\begin{array}{rl} 7 & \text{seven} \\ \underline{-4} & \text{minus four} \\ 3 & \text{equals three} \end{array}$$

Order in Subtraction

Subtraction problems with whole numbers can only be done in one order. You must subtract the smaller number from the larger number.

Look at these examples:

$8 - 2 = 6$ In problems that are written this way, subtract the second number from the first. In this example, you subtract the 2 from the 8.

$9 - 4 = 5$ In this example, you subtract the 4 from the 9.

$$\begin{array}{r} 8 \\ \underline{-2} \\ 6 \end{array}$$

In problems that are written this way, subtract the bottom number from the top. In this example you subtract 2 from 8.

$$\begin{array}{r} 9 \\ \underline{-4} \\ 5 \end{array}$$

In this example, you subtract 4 from 9.

PRACTICE

Read each example.

1. $9 - 4 = 5$

2. $8 - 1 = 7$

3. $6 - 5 = 1$

4. $7 - 6 =$

5. $\begin{array}{r} 9 \\ -7 \\ \hline \end{array}$

6. $\begin{array}{r} 2 \\ -1 \\ \hline \end{array}$

7. $\begin{array}{r} 5 \\ -3 \\ \hline \end{array}$

8. $\begin{array}{r} 6 \\ -4 \\ \hline \end{array}$

9. $\begin{array}{r} 46 \\ -25 \\ \hline \end{array}$

10. $\begin{array}{r} 79 \\ -49 \\ \hline \end{array}$

11. $\begin{array}{r} 25 \\ -12 \\ \hline \end{array}$

12. $\begin{array}{r} 63 \\ -32 \\ \hline \end{array}$

Using a Number Line to Subtract

A number line can help you subtract.

Look at the example and the number line below.

Example: $9 - 3 = 6$

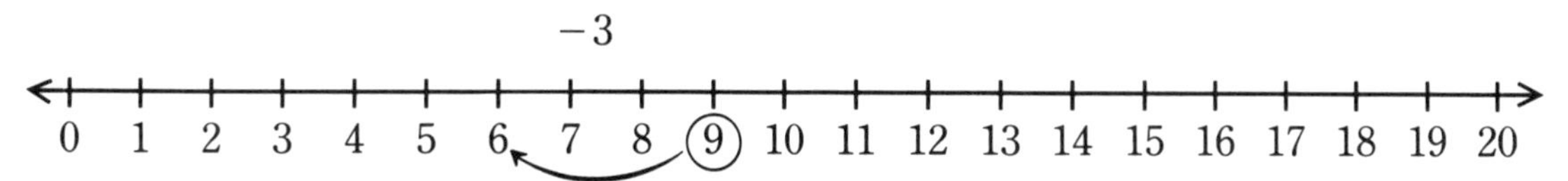

Use the number line to subtract. Follow the steps.

Step 1: You are going to subtract 3 from 9. Find 9 on the number line.

On a number line: Go *right* to add, go *left* to subtract.

Step 2: To subtract 3, move three numbers to the *left*.
You are at 6.
The number line shows that $9 - 3 = 6$.

PRACTICE

A. Copy the problems below on a separate sheet of paper. Use the number line to help you find the differences.

1. 9 − 1 =
2. 6 − 3 =
3. 4 − 3 =
4. 8 − 7 =
5. 3 − 2 =
6. 8 − 6 =
7. 9 − 6 =
8. 7 − 1 =
9. 7 − 5 =
10. 5 − 4 =
11. 4 − 2 =
12. 6 − 5 =
13. 7 − 3 =
14. 9 − 2 =
15. 7 − 6 =
16. 8 − 3 =
17. $\begin{array}{r} 8 \\ -2 \\ \hline \end{array}$
18. $\begin{array}{r} 8 \\ -4 \\ \hline \end{array}$
19. $\begin{array}{r} 9 \\ -3 \\ \hline \end{array}$
20. $\begin{array}{r} 4 \\ -1 \\ \hline \end{array}$

B. Copy the problems below on a separate sheet of paper. Find the differences.

1. $\begin{array}{r} 7 \\ -3 \\ \hline \end{array}$
2. $\begin{array}{r} 5 \\ -4 \\ \hline \end{array}$
3. $\begin{array}{r} 9 \\ -6 \\ \hline \end{array}$
4. $\begin{array}{r} 6 \\ -4 \\ \hline \end{array}$
5. $\begin{array}{r} 17 \\ -8 \\ \hline \end{array}$
6. $\begin{array}{r} 14 \\ -5 \\ \hline \end{array}$
7. $\begin{array}{r} 13 \\ -4 \\ \hline \end{array}$
8. $\begin{array}{r} 12 \\ -7 \\ \hline \end{array}$
9. $\begin{array}{r} 8 \\ -3 \\ \hline \end{array}$
10. $\begin{array}{r} 9 \\ -7 \\ \hline \end{array}$
11. $\begin{array}{r} 12 \\ -4 \\ \hline \end{array}$
12. $\begin{array}{r} 16 \\ -9 \\ \hline \end{array}$
13. $\begin{array}{r} 12 \\ -7 \\ \hline \end{array}$
14. $\begin{array}{r} 16 \\ -8 \\ \hline \end{array}$
15. $\begin{array}{r} 15 \\ -6 \\ \hline \end{array}$
16. $\begin{array}{r} 14 \\ -8 \\ \hline \end{array}$
17. $\begin{array}{r} 18 \\ -9 \\ \hline \end{array}$
18. $\begin{array}{r} 13 \\ -5 \\ \hline \end{array}$
19. $\begin{array}{r} 14 \\ -8 \\ \hline \end{array}$
20. $\begin{array}{r} 13 \\ -6 \\ \hline \end{array}$
21. $\begin{array}{r} 5 \\ -2 \\ \hline \end{array}$
22. $\begin{array}{r} 2 \\ -1 \\ \hline \end{array}$
23. $\begin{array}{r} 6 \\ -2 \\ \hline \end{array}$
24. $\begin{array}{r} 7 \\ -6 \\ \hline \end{array}$
25. $\begin{array}{r} 4 \\ -2 \\ \hline \end{array}$

Subtracting Zero

When you subtract zero from any whole number, the number stays the same. See these facts for yourself. Use the number line to prove that $9 - 0 = 9$.
Read these examples:

$9 - 0 = 9$ $21 - 0 = 21$ $682 - 0 = 682$
$3{,}592 - 0 = 3{,}592$

Subtracting a Number from Itself

When you subtract a number from itself, the answer is zero.

$9 - 9 = 0$ $21 - 21 = 0$ $682 - 682 = 0$
$3{,}592 - 3{,}592 = 0$

PRACTICE

Copy the problems below on a separate sheet of paper. Find the differences.

1. $3 - 0 =$

2. $7 - 7 =$

3. $16 - 16 =$

4. $17 - 0 =$

5. $10 - 10 =$

6. $10 - 0 =$

7. $13 - 0 =$

8. $412 - 412 =$

3.3 Subtracting Larger Numbers

Find the difference. $\begin{array}{r} 968 \\ -625 \\ \hline \end{array}$ Follow the steps below.

Step 1: Subtract the digits in the ones place.

Step 2: Subtract the digits in the tens place.

Step 3: Subtract the digits in the hundreds place.

Hundreds	Tens	Ones
9	6	8
−6	2	5
3	4	3

PRACTICE

A. Copy the problems below on a separate sheet of paper. Find the differences.

1. $\begin{array}{r} 30 \\ -10 \\ \hline \end{array}$
2. $\begin{array}{r} 67 \\ -41 \\ \hline \end{array}$
3. $\begin{array}{r} 71 \\ -30 \\ \hline \end{array}$
4. $\begin{array}{r} 76 \\ -45 \\ \hline \end{array}$
5. $\begin{array}{r} 95 \\ -3 \\ \hline \end{array}$
6. $\begin{array}{r} 85 \\ -52 \\ \hline \end{array}$
7. $\begin{array}{r} 54 \\ -30 \\ \hline \end{array}$
8. $\begin{array}{r} 47 \\ -12 \\ \hline \end{array}$
9. $\begin{array}{r} 66 \\ -10 \\ \hline \end{array}$
10. $\begin{array}{r} 99 \\ -7 \\ \hline \end{array}$
11. $\begin{array}{r} 64 \\ -23 \\ \hline \end{array}$
12. $\begin{array}{r} 59 \\ -43 \\ \hline \end{array}$
13. $\begin{array}{r} 95 \\ -40 \\ \hline \end{array}$
14. $\begin{array}{r} 73 \\ -52 \\ \hline \end{array}$
15. $\begin{array}{r} 87 \\ -25 \\ \hline \end{array}$

B. Copy the problems below on a separate sheet of paper. Find the differences.

1. $\begin{array}{r} 819 \\ -705 \\ \hline \end{array}$
2. $\begin{array}{r} 597 \\ -444 \\ \hline \end{array}$
3. $\begin{array}{r} 678 \\ -251 \\ \hline \end{array}$
4. $\begin{array}{r} 956 \\ -542 \\ \hline \end{array}$
5. $\begin{array}{r} 9{,}568 \\ -5{,}420 \\ \hline \end{array}$
6. $\begin{array}{r} 8{,}428 \\ -7{,}105 \\ \hline \end{array}$
7. $\begin{array}{r} 7{,}868 \\ -3{,}641 \\ \hline \end{array}$
8. $\begin{array}{r} 6{,}598 \\ -1{,}347 \\ \hline \end{array}$
9. $\begin{array}{r} 5{,}326 \\ -2{,}213 \\ \hline \end{array}$
10. $\begin{array}{r} 9{,}469 \\ -3{,}000 \\ \hline \end{array}$

3.4 Checking Subtraction

The answers to subtraction problems may be checked by adding. Check the answer to this subtraction problem. Follow the steps.

$$\begin{array}{r} 596 \\ -234 \\ \hline 362 \end{array}$$

Step 1: Add the number you have subtracted and the difference.

$$\begin{array}{r} 234 \\ +362 \\ \hline 596 \end{array}$$

Step 2: Compare the sum with the top number in the subtraction problem. If they are the same, your answer to the subtraction problem is correct.

PRACTICE

A. Copy these subtraction problems on a separate sheet of paper. Check the answer to each problem carefully. Then circle the problems with the wrong answers.

1.	**3.**	**5.**	**7.**	**9.**
547	978	8,658	6,739	9,658
−205	−763	−6,428	−5,501	−3,244
342	135	2,230	1,138	6,514

2.	**4.**	**6.**	**8.**	**10.**
987	896	9,457	6,698	7,378
−526	−583	−7,233	−4,245	−3,164
461	313	2,224	4,432	4,214

B. Copy the problems below on a separate sheet of paper. Find the differences. Check your answers by adding.

1.	**2.**	**3.**	**4.**	**5.**
931	846	7,759	8,674	8,797
−320	−632	−5,134	−6,432	−7,450

3.5 Renaming for Subtraction

Renaming to Show More Ones

You have learned that 24 can be renamed this way: two tens and four ones.

24 = 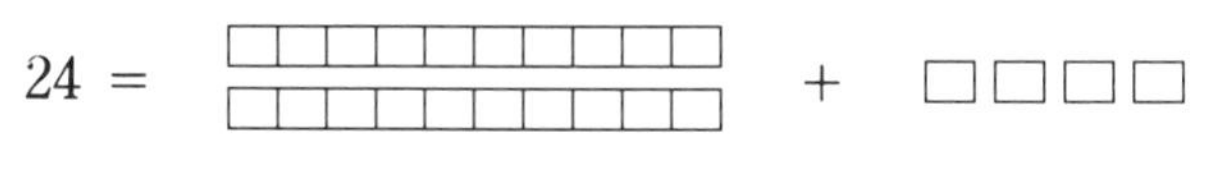

24 = two tens + four ones

The number 24 can also be renamed to show fewer tens and more ones.

There are 10 ones in each ten. 10 ones + 4 ones = 14 ones.

24 = 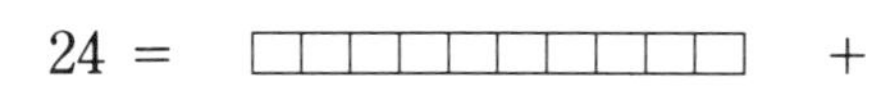 +

24 = one ten + fourteen ones

PRACTICE

A. Copy the number sentences on a separate sheet of paper. Rename the numbers in color to show one less ten and more ones. The first one has been done for you.

1. 39 can be renamed two tens and nineteen ones.
2. 54 can be renamed ______ tens and ______ ones.
3. 82 can be renamed ______ tens and ______ ones.
4. 73 can be renamed ______ tens and ______ ones.
5. 65 can be renamed ______ tens and ______ ones.
6. 47 can be renamed ______ tens and ______ ones.
7. 96 can be renamed ______ tens and ______ ones.
8. 22 can be renamed ______ tens and ______ ones.

Renaming to Show More Tens

You can rename 436 to show one less hundred and more tens.

There are 10 tens in each hundred. 10 tens + 3 tens = 13 tens.

436 =

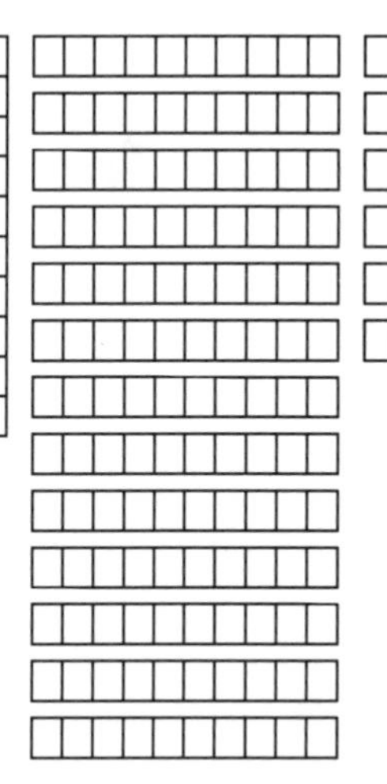

436 = three hundreds + thirteen tens + six ones

Renaming to Show More Hundreds

You can rename numbers to show one less thousand and more hundreds. There are ten hundreds in each thousand. When you rename to show more hundreds, you take ten hundreds and add them to the hundreds place.

thousands	hundreds	tens	ones	→	thousands	hundreds	tens	ones
6	4	2	1		5	14	2	1
					~~6~~	~~4~~		

PRACTICE

B. Copy the number sentences on a separate sheet of paper. Rename the numbers in color. Show one less hundred and more tens. Write the correct number words in the blanks. The first one has been done for you.

1. 352 can be renamed two hundreds + fifteen tens + two ones.
2. 694 can be renamed _______ hundreds + _______ tens + _______ ones.
3. 837 can be renamed _______ hundreds + _______ tens + _______ ones.
4. 259 can be renamed _______ hundreds + _______ tens + _______ ones.
5. 470 can be renamed _______ hundreds + _______ tens + _______ ones.
6. 531 can be renamed _______ hundreds + _______ tens + _______ ones.

C. Copy the number sentences on a separate sheet of paper. Rename the numbers in color. Show one less thousand and more hundreds. Write the correct number words in the blanks. The first one has been done for you.

1. 5,237 can be renamed four thousands, twelve hundreds, three tens, and seven ones.
2. 2,146 can be renamed _______ thousands, _______ hundreds, _______ tens, and _______ ones.
3. 8,629 can be renamed _______ thousands, _______ hundreds, _______ tens, and _______ ones.
4. 3,560 can be renamed _______ thousands, _______ hundreds, _______ tens, and _______ ones.
5. 9,804 can be renamed _______ thousands, _______ hundreds, _______ tens, and _______ ones.
6. 4,191 can be renamed _______ thousands, _______ hundreds, _______ tens, and _______ ones.

3.6 Subtraction with One Renaming

Renaming to Show More Ones

In some problems you must rename to show more ones before you can subtract. Follow the steps to subtract this problem:

$$\begin{array}{r} 52 \\ -17 \\ \hline \end{array}$$

Step 1: Rename to show more ones.
52 = 4 tens + 12 ones

$$\begin{array}{rr} 4 & 12 \\ \not{5} & \not{2} \\ -1 & 7 \\ \hline \end{array}$$

Step 2: Subtract the digits in the ones place.
12 − 7 = 5

$$\begin{array}{rr} 4 & 12 \\ \not{5} & \not{2} \\ -1 & 7 \\ \hline & 5 \end{array}$$

Step 3: Subtract the digits in the tens place.
4 − 1 = 3

$$\begin{array}{rr} 4 & 12 \\ \not{5} & \not{2} \\ -1 & 7 \\ \hline 3 & 5 \end{array}$$

PRACTICE

A. Copy the problems below on a separate sheet of paper. Find the differences. Remember to rename to show more ones.

1. 83 − 16	**4.** 92 − 37	**7.** 61 − 39	**10.** 87 − 19	**13.** 32 − 23
2. 74 − 45	**5.** 51 − 32	**8.** 95 − 7	**11.** 57 − 18	**14.** 93 − 57
3. 68 − 39	**6.** 77 − 48	**9.** 76 − 58	**12.** 64 − 36	**15.** 82 − 45

Renaming to Show More Tens or Hundreds

In some problems you must rename to show more tens or hundreds before you can subtract.

Follow the steps to solve this problem.

$$\begin{array}{r} 638 \\ -172 \\ \hline \end{array}$$

Step 1: Subtract the digits in the ones place.

Step 2: Rename to show more tens.

Step 3: Subtract the digits in the tens and hundreds places.

$$\begin{array}{rrr} 5 & 13 & \\ \not{6} & \not{3} & 8 \\ -1 & 7 & 2 \\ \hline 4 & 6 & 6 \end{array}$$

Now follow the steps to solve this problem.

$$\begin{array}{r} 8574 \\ -6831 \\ \hline \end{array}$$

Step 1: Subtract the digits in the ones and tens places.

Step 2: Rename to show more hundreds.

Step 3: Subtract the hundreds and thousands places.

$$\begin{array}{rrrr} 7 & 15 & & \\ \not{8} & \not{5} & 7 & 4 \\ -6 & 8 & 3 & 1 \\ \hline 1 & 7 & 4 & 3 \end{array}$$

PRACTICE

B. Copy the problems below on a separate sheet of paper. Find the differences. Rename whenever you have to.

1. 419 − 295

2. 518 − 247

3. 622 − 371

4. 836 − 545

5. 8,362 − 5,450

6. 8,428 − 7,195

7. 5,724 − 2,800

8. 9,856 − 3,009

9. 685 − 392

10. 867 − 359

11. 3,622 − 1,715

12. 8,533 − 6,604

3.7 Subtracting with More Than One Renaming

In some problems you need to rename more than once.
Follow the steps to solve this problem.

$$\begin{array}{r} 731 \\ -298 \\ \hline \end{array}$$

Step 1: Rename to show more ones.

Step 2. Subtract the digits in the ones place.

$$\begin{array}{rrr} & 2 & 11 \\ 7 & \cancel{3} & \cancel{1} \\ -2 & 9 & 8 \\ \hline & & 3 \end{array}$$

Step 3: Rename to show more tens.

Step 4: Subtract the digits in the tens and hundreds place.

$$\begin{array}{rrr} & 12 & \\ 6 & \cancel{2} & 11 \\ \cancel{7} & \cancel{3} & \cancel{1} \\ -2 & 9 & 8 \\ \hline 4 & 3 & 3 \end{array}$$

PRACTICE

Copy the problems below on a separate sheet of paper. Find the differences. Rename whenever you have to.

1. 523 − 165	**6.** 927 − 247	**11.** 4,726 − 2,853	**16.** 7,378 − 4,889	**21.** 858 − 299
2. 714 − 358	**7.** 652 − 462	**12.** 8,614 − 6,796	**17.** 7,753 − 4,846	**22.** 723 − 446
3. 395 − 207	**8.** 739 − 545	**13.** 5,239 − 3,655	**18.** 7,582 − 5,619	**23.** 3,535 − 1,287
4. 821 − 473	**9.** 9,532 − 5,860	**14.** 9,268 − 1,749	**19.** 6,492 − 1,657	**24.** 6,225 − 4,877
5. 871 − 281	**10.** 6,423 − 3,186	**15.** 8,451 − 5,953	**20.** 9,673 − 5,794	**25.** 7,325 − 4,717

3.8 Renaming with Zeros

Solve this subtraction problem by following the steps.

$$\begin{array}{r} 8{,}000 \\ -5{,}624 \\ \hline \end{array}$$

Step 1: Rename to show more hundreds.

```
 7  10
 8̸  0̸  0  0
-5  6   2  4
```

Step 2: Rename to show more tens.

```
     9
 7  1̸0̸ 10
 8̸  0̸  0̸  0
-5  6   2  4
```

Step 3: Rename to show more ones.

```
     9   9
 7  1̸0̸  1̸0̸  10
 8̸  0̸   0̸   0̸
-5  6    2   4
```

Step 4: Subtract.

```
 2  3    7   6
```

PRACTICE

Copy the problems below on a separate sheet of paper.
Find the differences. Rename whenever you have to.

1. 600 − 276	**5.** 701 − 364	**9.** 9,000 − 5,920	**13.** 7,040 − 5,467	**17.** 6,500 − 1,932
2. 802 − 629	**6.** 453 − 206	**10.** 5,841 − 1,972	**14.** 6,001 − 4,635	**18.** 8,020 − 6,430
3. 400 − 103	**7.** 807 − 569	**11.** 7,006 − 2,000	**15.** 8,000 − 2,974	**19.** 9,200 − 1,831
4. 759 − 378	**8.** 647 − 49	**12.** 8,539 − 6,584	**16.** 9,079 − 4,389	**20.** 7,685 − 4,787

3.9 Estimating and Thinking

Estimating can help you subtract correctly. It will quickly give you an answer that is close to the exact one. Use rounded numbers to estimate.

Reviewing Rounding Numbers

	Rounding to the Nearest		
	Ten	Hundred	Thousand
Choose the place you are rounding to.	73	581	6,543
Ask yourself if the digit to the right is 5 or more.	73	581	6,543
If it is, add 1 to the place you are rounding to.		+1	+1
If it isn't, make no change.			
Replace the digits to the right with zeros.	70	600	7,000

Follow the steps below to help you estimate the answer to the example.

Step 1: Round the numbers to the nearest hundred.

Step 2: Subtract the rounded numbers.

Step 3: Subtract the numbers in the example. Compare the differences.

Exact	Rounded
778	800
−217	−200
561	600

PRACTICE

Copy the problems below on a separate sheet of paper. Round the numbers to the nearest hundred. Estimate the answers. Then subtract to find the exact differences.

1. 735 − 230
2. 926 − 864
3. 649 − 483
4. 678 − 289
5. 465 − 192
6. 542 − 163
7. 857 − 509
8. 864 − 508
9. 207 − 126
10. 945 − 362
11. 756 − 422
12. 593 − 229
13. 649 − 355
14. 924 − 666
15. 858 − 399

3.10 Solving Word Problems in Subtraction

Steps in Solving Word Problems

Read this problem. Follow the steps below to solve it.

Kim and Paul hiked 9 miles on Tuesday. They hiked 12 miles on Wednesday. How many more miles did they hike on Wednesday than on Tuesday?

Step 1: Read the problem.

Step 2: Learn what you must find out.

How many more miles did Kim and Paul hike on Wednesday? Clue Words: how many more

Step 3: Notice the clue words. Clue words give you the clue to solving the problem. *How many more* tells you to find the difference between two numbers. To solve this problem, you should *subtract.*

Step 4: Write down the numbers. Which is the larger number? Always subtract the smaller number from the larger number.

Larger Number: 12
Number to Subtract: 9

Step 5: Subtract to find the difference.

$$\begin{array}{r} 12 \\ -\ 9 \\ \hline 3 \end{array}$$

Step 6: Check the answer. Ask yourself if it makes sense.

Clue Words

Words and groups of words like these give you a clue. They tell you to find the difference between numbers. They tell you how to subtract.

how many more *how many less* *how much more* *remain* *fewer*
how much larger *how much shorter* *left* *difference* *greater*

Reminder: Not all problems have clue words.

PROBLEMS TO SOLVE

Solve the problems below on a separate sheet of paper.

All of these problems can be solved by using the steps on page 49.

1. Nat's car cost $8,739. Lori's car cost $6,952. How much more did Nat's car cost?
2. Kay saved $280. Meg saved $155. How much less did Meg save?
3. There are 1,097 students at Wilson High School. At Damato High School, there are 2,052 students. How many fewer students did Wilson have?
4. The show was on for two nights. The first night 397 people came. The second night 482 came. How many more people came on the second night?
5. The monthly rent for apartment 1A is $364. The rent for apartment 1B is $485. What is the difference in the rents?
6. Norma makes $320 a week. She spends $285 a week. How much does she have left?
7. Cal had 650 boxes to pack. By noon he had packed 287. How many more did he still have to pack?
8. Jan bought a used car. It had been driven 27,895 miles. After Jan had the car for a year there were 36,781 miles on it. How many miles did Jan drive that year?
9. The television set costs $398. It's on sale for $325. How much less does it cost on sale?
10. Jeff paid $329 for a new camera. Linda paid $279 for the same camera. How much less did Linda pay than Jeff?

3.11 Add or Subtract

Sometimes you must decide whether to add or subtract in order to solve a problem. The clue words can help you decide. Read them carefully.

Before you begin, read the clue word boxes on pages 29 and 49.

Marla drives to work. Driving costs her $532 a year. Joel takes the bus to work. Bus fare cost Joel $366 a year. How much more does Marla spend?

To solve the problem, follow the steps below.

Step 1: Read the problem.

Step 2: Learn what you must find out.

Step 3: Notice the clue words.

Step 4: Decide whether to add or subtract.

Step 5: Solve the problem.

Step 6: Check the answer. Ask yourself if it makes sense.

PROBLEMS TO SOLVE

Read each problem below. Decide whether to add or subtract. Then solve the problem on a separate sheet of paper.

1. Tony paid $3,762 for a car. He spent $65 for a new tire. His car insurance cost $402. How much did he spend in all?
2. Diane spent $320 last week. She spent $247 this week. How much less did she spend this week?
3. Ramon bought 600 boxes of cards. He sold 439. How many has he left to sell?
4. Barbara bought a new computer for $1,500. She also bought a printer for $479 and a box of disks for $49. How much did she spend in all?

MATHEMATICS IN YOUR LIFE:
Weekly Expenses

Larry makes $50 a week at his part-time job. He wants to estimate how much money he might save each week. So he keeps a record of his expenses for a week.

Day	Expense	Amount
Sunday	movie and food	$7.00
Monday	lunch	1.25
	bus fare	1.00
	music tape	5.50
Tuesday	lunch	.95
	bus fare	1.00
	school supplies	2.75
Wednesday	lunch	1.55
	bus fare	1.00
Thursday	lunch	1.35
	bus fare	1.00
	clothing	8.35
Friday	lunch	1.05
	bus fare	1.00
	food	2.00
Saturday	dance ticket	4.00

1. How much money did Larry spend that week?
2. How much money did he have left?

Every weekday Larry has to spend money for two things. Estimating how much he must spend for these things will help him plan ahead.

3. What two expenses does Larry have every weekday?
4. About how much does he spend in a week on these two expenses?
5. What amount does Larry spend on bus fare for the week?

3.12 Using Your Calculator: Subtraction

A calculator makes subtracting large numbers easy. Follow the steps below to subtract the numbers.

Step 1. Turn on your calculator. Find the [−] sign. Find the [=] sign.

Step 2: Press [9] [0] [4] [6]

Step 3: Press [−]

Step 4: Press [4] [2] [9] [8]

Step 5: Press [=]

Step 6: Your answer should be: 4,748

PRACTICE

A. Use a calculator to find the difference between the numbers in each pair. Write your answers on a separate sheet of paper.

1. 8,300 5,619 **3.** 3,572 1,679 **5.** 6,000 4,923

2. 9,603 8,286 **4.** 7,505 2,396 **6.** 4,803 2,957

B. Use a calculator to find the difference for each problem. Write your answers on a separate sheet of paper.

1. $\begin{array}{r} 6{,}302 \\ -4{,}807 \\ \hline \end{array}$ **2.** $\begin{array}{r} 7{,}457 \\ -3{,}998 \\ \hline \end{array}$ **3.** $\begin{array}{r} 5{,}200 \\ -3{,}866 \\ \hline \end{array}$ **4.** $\begin{array}{r} 9{,}003 \\ -6{,}756 \\ \hline \end{array}$

CHAPTER REVIEW

CHAPTER SUMMARY

■ Subtract	To take one number away from another; to find the difference between two numbers.	$9 - 4 = 5$
■ Subtracting 0	When you subtract 0 from a number, the number stays the same.	$9 - 0 = 9$
■ Subtracting a Number from Itself	When you subtract a number from itself, the answer is zero.	$9 - 9 = 0$
■ Subtracting Large Numbers	To subtract larger numbers: start with the digits in the ones place. Then subtract one place at a time.	$\begin{array}{r} 9{,}642 \\ -3{,}268 \\ \hline 5{,}374 \end{array}$
■ Renaming	Sometimes you cannot subtract the digits in a column. In that case, you must rename to find the difference.	$\begin{array}{r} 7{,}510 \\ -5{,}637 \\ \hline 1{,}873 \end{array}$
■ Estimating	Estimating gives you a quick idea of what an answer will be. Rounding numbers helps you estimate.	
■ Solving Subtraction Problems Step-by-Step	1) Read the problem. 2) Learn what you must find out. 3) Find the numbers and the clue. 4) Decide whether to add or subtract. 4) Subtract to find the difference.	

CHAPTER REVIEW

CHAPTER QUIZ

A. Copy the problems on a separate sheet of paper. Find the differences.

1. $\begin{array}{r} 953 \\ -341 \\ \hline \end{array}$	**4.** $\begin{array}{r} 638 \\ -495 \\ \hline \end{array}$	**7.** $\begin{array}{r} 411 \\ -307 \\ \hline \end{array}$	**10.** $\begin{array}{r} 808 \\ -527 \\ \hline \end{array}$
2. $\begin{array}{r} 700 \\ -469 \\ \hline \end{array}$	**5.** $\begin{array}{r} 824 \\ -753 \\ \hline \end{array}$	**8.** $\begin{array}{r} 625 \\ -263 \\ \hline \end{array}$	**11.** $\begin{array}{r} 930 \\ -714 \\ \hline \end{array}$
3. $\begin{array}{r} 8,695 \\ -2,694 \\ \hline \end{array}$	**6.** $\begin{array}{r} 7,593 \\ -6,895 \\ \hline \end{array}$	**9.** $\begin{array}{r} 8,732 \\ -\quad 641 \\ \hline \end{array}$	**12.** $\begin{array}{r} 5,720 \\ -\quad 872 \\ \hline \end{array}$

B. Developing Thinking Skills

Read the problem. Then answer the questions on a separate piece of paper.

The book was 176 pages long. Cory read 89 pages. How many pages did Cory have left to read?

1. What is the clue word or words?
2. Should you add or subtract to solve this problem?
3. What are the numbers in the problem?
4. What is the answer to the problem?

REVIEWING VOCABULARY

Copy the sentences on a separate sheet of paper. Fill in the blanks with the correct word from the box.

rename	subtract	difference

1. The ________ is the amount by which one number is larger or smaller than another.
2. You must ________ to find out how much is left.
3. Sometimes you cannot subtract the digits in a problem unless you ________.

MULTIPLYING WHOLE NUMBERS

4

The thunderstorm pictured here dropped a large amount of rain on this city in a very brief time. Suppose a city received about 15 inches of rain each month. How can you learn how much rain that city received in a year? That's right—you multiply. (15×12)

Chapter Learning Objectives

1. Multiply whole numbers
2. Multiply by zero
3. Multiply by one
4. Multiply by tens, hundreds, and thousands
5. Multiply larger numbers
6. Multiply with renaming
7. Estimate answers
8. Use multiplication to solve word problems

WORDS TO KNOW

■ multiply	to add a number to itself one or more times $2 + 2 + 2 + 2 = 8 \qquad 2 \times 4 = 8$
■ multiplication	a quick way to add
■ product	the final answer to a multiplication problem
■ partial product	the number obtained by multiplying a number by only one digit of a number
■ factor	one of the numbers that must be multiplied to obtain a product
■ result	the answer to a problem

4.1 What Is Multiplication?

Multiplication is the process of adding a number to itself one or more times.

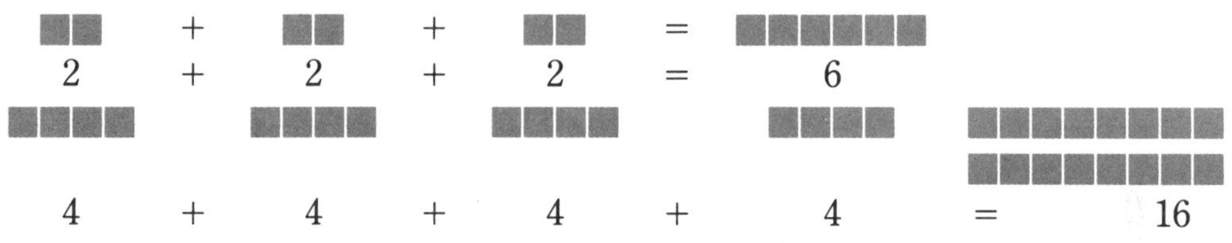

Multiplication is a faster way to get the same **result.**

3	×	2	=	6
three	times	two	equals	six

4	×	4	=	16
four	times	four	equals	sixteen

In this chapter you will learn more about multiplication. You will also learn when and how to use multiplication to solve word problems.

4.2 Basic Multiplication

Reading and Writing Multiplication Problems

Multiplication problems may be written in two ways. Read the examples below.

3	×	3	=	9
three	times	three	equals	nine

$$\begin{array}{rl} 3 & \text{three} \\ \times 3 & \text{times three} \\ \hline & \text{equals} \\ 9 & \text{nine} \end{array}$$

Each number being multiplied is called a **factor.** The result, or answer, is called the **product.**

$$\begin{array}{rl} 3 & \text{factor} \\ \underline{\times 4} & \text{factor} \\ 12 & \text{product} \end{array}$$

Order in Multiplication

You can **multiply** factors in any order. The product will not change.

Here are two groups of three. ■■■ ■■■ ■■■■■■

3 × 2 = 6

Here are three groups of two. ■■ ■■ ■■ ■■■■■■

2 × 3 = 6

PRACTICE

A. Read each example.

1. $\begin{array}{r} 6 \\ \underline{\times 2} \\ 12 \end{array}$

2. $\begin{array}{r} 5 \\ \underline{\times 4} \\ 20 \end{array}$

3. $\begin{array}{r} 7 \\ \underline{\times 8} \\ 56 \end{array}$

4. $\begin{array}{r} 9 \\ \underline{\times 5} \\ 45 \end{array}$

5. $\begin{array}{r} 7 \\ \underline{\times 7} \\ 49 \end{array}$

6. $\begin{array}{r} 3 \\ \underline{\times 6} \\ 18 \end{array}$

7. $\begin{array}{r} 8 \\ \underline{\times 4} \\ 32 \end{array}$

8. $\begin{array}{r} 6 \\ \underline{\times 7} \\ 42 \end{array}$

9. $\begin{array}{r} 5 \\ \underline{\times 5} \\ 25 \end{array}$

A Multiplication Chart

Remembering multiplication facts makes multiplying quicker and easier. One way to learn multiplication is by studying a multiplication chart.

Look at the multiplication chart below. It shows how to multiply numbers through the number 36. It is easy to use the chart. Suppose you wanted to multiply 5×6. Follow the steps.

Step 1: Find the number 5 in the first column going down.

Step 2: Then find the number 6 in the top row going across.

Step 3: Move across from the 5 and down from the 6 to find where the row meets the column.

Step 4: The box where the row meets the column shows the product of the two numbers.

$5 \times 6 = 30$

Study the chart below. Practice multiplying the numbers in the first column going down by the numbers going across in the top row.

1	2	3	4	5	6	7	8	9
2	4	6	8	10	12	14	16	18
3	6	9	12	15	18	21	24	27
4	8	12	16	20	24	28	32	36
5	10	15	20	25	30	35	40	45
6	12	18	24	30	36	42	48	54
7	14	21	28	35	42	49	56	63
8	16	24	32	40	48	56	64	72
9	18	27	36	45	54	63	72	81

PRACTICE

B. Copy the problems below on a separate sheet of paper. Find the products. If you need help, turn to the multiplication chart on page 348 in the back of the book.

1. $4 \times 2 =$
2. $2 \times 4 =$
3. $5 \times 3 =$
4. $3 \times 5 =$
5. $6 \times 4 =$
6. $4 \times 6 =$
7. $5 \times 8 =$
8. $8 \times 5 =$
9. $6 \times 2 =$
10. $2 \times 6 =$
11. $2 \times 9 =$
12. $9 \times 2 =$
13. $5 \times 4 =$
14. $4 \times 5 =$
15. $4 \times 8 =$
16. $8 \times 4 =$
17. $6 \times 9 =$
18. $9 \times 6 =$
19. $6 \times 5 =$
20. $5 \times 6 =$
21. $7 \times 8 =$
22. $8 \times 7 =$
23. $9 \times 4 =$
24. $4 \times 9 =$
25. $5 \times 7 =$
26. $7 \times 5 =$
27. $8 \times 9 =$
28. $9 \times 8 =$

C. Copy the problems below on a separate sheet of paper. Find the products.

1. 3×8
2. 8×4
3. 6×7
4. 7×5
5. 2×5
6. 6×3
7. 5×9
8. 9×6
9. 9×3
10. 9×7
11. 8×6
12. 7×8
13. 6×7
14. 6×6
15. 9×4
16. 9×5
17. 8×7
18. 7×9
19. 8×8
20. 9×9
21. 4×9
22. 7×4
23. 6×5
24. 3×7
25. 9×2

4.3 Multiplying with Zero and One

When you multiply any whole number by zero, the answer is zero. When you multiply zero by any whole number, the answer is zero.

Examples: $6 \times 0 = 0$ $0 \times 6 = 0$

When you multiply any whole number by one, the number stays the same.

Examples: $6 \times 1 = 6$ $1 \times 6 = 6$
$9 \times 1 = 9$ $1 \times 9 = 9$

PRACTICE

A. Copy the problems below on a separate sheet of paper. Find the products.

1. $3 \times 0 =$
2. $1 \times 1 =$
3. $1 \times 0 =$
4. $6 \times 1 =$
5. $7 \times 0 =$
6. $8 \times 0 =$
7. $0 \times 7 =$
8. $0 \times 0 =$
9. $5 \times 1 =$
10. $1 \times 9 =$
11. $8 \times 1 =$
12. $0 \times 1 =$

B. Copy the problems below on a separate sheet of paper. Find the products.

1. 6×1
2. 9×3
3. 0×7
4. 9×5
5. 1×1
6. 9×9
7. 2×0
8. 0×5
9. 8×1
10. 7×8
11. 5×9
12. 7×7
13. 8×7
14. 1×9
15. 3×8
16. 7×3
17. 8×0
18. 6×5
19. 0×4
20. 9×2

4.4 Multiplying by Larger Numbers

Two-Digit Numbers

Look at this example:

$$\begin{array}{r} 72 \\ \times 23 \\ \hline \end{array}$$

It tells you to multiply 72 by 23.

Follow these steps.

Step 1: Multiply 72 by the digit in the *ones* place.

$2 \times 3 = 6$
$7 \times 3 = 21$

$$\begin{array}{r} 72 \\ \times 23 \\ \hline 216 \end{array} \text{ partial product}$$

Step 2: Multiply 72 by the digit in the *tens* place. Write the second partial product one place to the left in the *tens* place.

$2 \times 2 \text{ tens} = 4 \text{ tens}$
$7 \times 2 \text{ tens} = 14 \text{ tens}$

$$\begin{array}{rl} 72 & \\ \times 23 & \\ \hline 216 & \text{partial product} \\ 144 & \text{partial product} \\ \hline 1656 & \text{product} \end{array}$$

Step 3: Add the partial products.

Remember: the **partial product** only gives part of an answer to a multiplication problem.

PRACTICE

A. Copy the problems below on a separate sheet of paper. Find the products.

1. $\begin{array}{r} 63 \\ \times 23 \\ \hline \end{array}$ **4.** $\begin{array}{r} 87 \\ \times 11 \\ \hline \end{array}$ **7.** $\begin{array}{r} 604 \\ \times 22 \\ \hline \end{array}$ **10.** $\begin{array}{r} 912 \\ \times 42 \\ \hline \end{array}$ **13.** $\begin{array}{r} 712 \\ \times 24 \\ \hline \end{array}$

2. $\begin{array}{r} 92 \\ \times 31 \\ \hline \end{array}$ **5.** $\begin{array}{r} 731 \\ \times 32 \\ \hline \end{array}$ **8.** $\begin{array}{r} 532 \\ \times 13 \\ \hline \end{array}$ **11.** $\begin{array}{r} 44 \\ \times 33 \\ \hline \end{array}$ **14.** $\begin{array}{r} 630 \\ \times 13 \\ \hline \end{array}$

3. $\begin{array}{r} 94 \\ \times 12 \\ \hline \end{array}$ **6.** $\begin{array}{r} 810 \\ \times 86 \\ \hline \end{array}$ **9.** $\begin{array}{r} 833 \\ \times 21 \\ \hline \end{array}$ **12.** $\begin{array}{r} 62 \\ \times 43 \\ \hline \end{array}$ **15.** $\begin{array}{r} 802 \\ \times 42 \\ \hline \end{array}$

Three-Digit Numbers

Follow the steps below to find the product of 623 × 123.

Step 1: Multiply 623 by the digit in the *ones* place.

$3 \times 3 = 9$
$2 \times 3 = 6$
$6 \times 3 = 18$

```
   623
 × 123
  1869  partial product
```

Step 2: Multiply 623 by the number in the *tens* place. Write the second partial product one place to the left in the *tens* place.

$3 \times 2 \text{ tens} = 6 \text{ tens}$
$2 \times 2 \text{ tens} = 4 \text{ tens}$
$6 \times 2 \text{ tens} = 12 \text{ tens}$

```
   623
 × 123
  1869  partial product
 1246   partial product
```

Step 3: Multiply 623 by the digit in the *hundreds* place. Write the third partial product one more place to the left in the *hundreds* place.

Step 4: Add the partial products.

```
    623
  × 123
   1869  partial product
  1246   partial product
  623    partial product
 76,629  product
```

PRACTICE

B. Copy the problems below on a separate sheet of paper. Find the products.

1.	**2.**	**3.**	**4.**	**5.**
943 × 121	731 × 332	811 × 679	823 × 323	701 × 968

C. On a separate sheet of paper rewrite the following multiplication problems so that they are easier to do. Then find the products. The first one has been rewritten for you.

1. 8,134 × 231 =
```
   8,134
 ×   231
```

2. 9,201 × 423 =

3. 7,423 × 213 =

4. 6,113 × 312 =

5. 8,120 × 342 =

6. 5,201 × 432 =

4.5 Multiplying with One Renaming

When the product in a place is larger than 9, renaming makes multiplication easy.

Look at this example:

$$\begin{array}{r} 75 \\ \times 6 \\ \hline \end{array}$$

It tells you to multiply 75 by 6.
You could multiply the problem this way:

$$\begin{array}{rr} & 75 \\ & \times\ 6 \\ \hline 5 \times 6 = & 30 \\ 70 \times 6 = & 420 \\ \hline & 450 \end{array}$$

But the problem is easier to solve when you rename.

Follow these steps.

Step 1: Multiply the digit in the ones place by 6.
5 × 6 = 30
Rename 30: 3 tens + 0 ones
Put the 0 in the ones place.
Put the 3 above the tens place

$$\begin{array}{r} 3\ \ \\ 75 \\ \times\ 6 \\ \hline 0 \end{array}$$

Step 2: Multiply the digit in the tens place by 6.
7 × 6 = 42
Add 42 + 3 = 45
Put the answer in the tens and hundreds places.

$$\begin{array}{r} 3\ \ \\ 75 \\ \times\ 6 \\ \hline 450 \end{array}$$ product

PRACTICE

Copy the problems below on a separate sheet of paper.
Find the products.

1. 67 × 9
2. 96 × 7
3. 95 × 7
4. 87 × 4
5. 78 × 8
6. 871 × 6
7. 681 × 5
8. 552 × 4
9. 817 × 6
10. 982 × 3

4.6 Multiplying with More Than One Renaming

Follow these steps.

Step 1: Multiply the digit in the ones place by 4.
$7 \times 4 = 28$
Rename 28: 2 tens + 8 ones
Put the 8 in the ones place.
Put the 2 above the tens place.

$$\begin{array}{r} 2 \\ 257 \\ \times \quad 4 \\ \hline 8 \end{array}$$

Step 2: Multiply the digit in the tens place by 4.
$5 \times 4 = 20$
Add $20 + 2 = 22$
Put 2 in the tens place.
Put 2 above the hundreds place.

$$\begin{array}{r} 22 \\ 257 \\ \times \quad 4 \\ \hline 28 \end{array}$$

Step 3: Multiply the digit in the hundreds place by 4.
$2 \times 4 = 8$
Add $8 + 2 = 10$
Put the product in the hundreds and thousands places.

$$\begin{array}{r} 22 \\ 257 \\ \times \quad 4 \\ \hline 1{,}028 \end{array}$$

PRACTICE

Write the problems below on a separate sheet of paper.
Find the products.

1. 576×97
2. 958×76
3. 784×89
4. 682×65
5. 857×36
6. 962×79
7. 873×48
8. 875×56
9. 569×94
10. 948×83
11. 843×847
12. 925×707
13. 692×634
14. 447×825
15. 564×628
16. 537×26
17. 482×35
18. 747×56
19. 625×324
20. 873×248

4.7 Multiplying by Tens, Hundreds, and Thousands

To multiply a number by 10, first write the number. Then write a zero to the right of the number.

$45 \times 10 = 450$

To multiply a number by 100, first write the number. Then write two 0's.

$62 \times 100 = 6{,}200$

To multiply a number by 1000, first write the number. Then write three 0's.

$95 \times 1000 = 95{,}000$

Did You Know?
A cow that gives a lot of milk can produce about 100 pints a day. That's 700 pints a week. That means that in one year that cow can produce enough milk (700 × 52 = 36,400 pints) to flood your living room!

PRACTICE

A. Copy the problems below on a separate sheet of paper. Find the products.

1. $37 \times 10 =$	**5.** $86 \times 100 =$	**9.** $59 \times 1{,}000 =$
2. $29 \times 100 =$	**6.** $5{,}279 \times 100 =$	**10.** $350 \times 10 =$
3. $596 \times 100 =$	**7.** $421 \times 1{,}000 =$	**11.** $140 \times 1{,}000 =$
4. $152 \times 1{,}000 =$	**8.** $3{,}715 \times 1{,}000 =$	**12.** $7{,}000 \times 1{,}000 =$

B. Copy the problems below on a separate sheet of paper. Find the products.

1. $\begin{array}{r} 756 \\ \times\ 10 \\ \hline \end{array}$

2. $\begin{array}{r} 87 \\ \times\ 10 \\ \hline \end{array}$

3. $\begin{array}{r} 259 \\ \times 100 \\ \hline \end{array}$

4. $\begin{array}{r} 7{,}356 \\ \times\ 100 \\ \hline \end{array}$

4.8 Multiplying by Numbers That Contain Zero

Look at this problem. There is a zero in the ones place. This means there are no ones.
You could multiply 83 by 0. But there is no need to. Any number multiplied by 0 equals 0. So just put a 0 in the ones place. Then multiply the rest of the numbers in the problem.

$$\begin{array}{r} 83 \\ \underline{\times 20} \\ 00 \\ \underline{1\,66} \\ 1{,}660 \end{array} \qquad \begin{array}{r} 83 \\ \underline{\times 20} \\ 1{,}660 \end{array}$$

Look at this problem. There is a zero in the tens place. This means there are no tens.
You could multiply 912 by 0 tens. But there is no need to. Just put a 0 in the tens place. Then multiply by the 3 in the hundreds place.

$$\begin{array}{r} 912 \\ \underline{\times 304} \\ 3\,648 \\ 0 \\ \underline{273\,6} \\ 277{,}248 \end{array} \qquad \begin{array}{r} 912 \\ \underline{\times 304} \\ 3\,648 \\ \underline{273\,60} \\ 277{,}248 \end{array}$$

PRACTICE

A. Copy the problems below on a separate sheet of paper. Find the products.

1. 498 × 70
2. 635 × 804
3. 537 × 420
4. 926 × 306
5. 829 × 470
6. 805 × 609
7. 793 × 50
8. 625 × 709
9. 970 × 260
10. 387 × 942
11. 369 × 549
12. 761 × 801
13. 580 × 409
14. 296 × 378
15. 807 × 903

B. Copy the problems below on a separate sheet of paper. Find the products.

1. 623 × 40 =
2. 29 × 60 =
3. 535 × 402 =
4. 753 × 250 =
5. 645 × 204 =
6. 85 × 50 =

4.9 Estimating and Thinking

When you need to estimate an answer, rounding numbers can be very helpful. Suppose you wanted to round the factors in these problems.

If you need help in rounding, turn back to pages 10–12.

Round both numbers to the nearest hundred.

$$\begin{array}{r} 392 \\ \times 587 \\ \hline \end{array} \qquad \begin{array}{r} 400 \\ \times 600 \\ \hline \end{array}$$

Round this number to the nearest thousand.
Round this number to the nearest hundred.

$$\begin{array}{r} 6,128 \\ \times \quad 762 \\ \hline \end{array} \qquad \begin{array}{r} 6,000 \\ \times \quad 800 \\ \hline \end{array}$$

PRACTICE

A. Round each number below on a separate sheet of paper. Round each two-digit number to the nearest ten. Round each three-digit number to the nearest hundred. Round each four-digit number to the nearest thousand.

1. 682
2. 5,439
3. 1,539
4. 375
5. 594
6. 8,425
7. 29
8. 4,631

B. Copy the problems below on a separate sheet of paper. First round the factors of each problem. Next multiply to estimate the answer. Then multiply to find the exact products. Round the numbers the same way you did in the above exercise.

1. 835×56
2. $2,592 \times 23$
3. $4,520 \times 893$
4. $6,920 \times 359$
5. $8,364 \times 7,428$
6. 408×46
7. 365×92
8. 198×639
9. 617×370
10. 852×509
11. $9,737 \times 951$
12. $5,364 \times 6,829$

4.10 Solving Multiplication Word Problems

Steps in Solving Word Problems

Read this problem. Follow the steps below to solve it.

Norm packs 16 boxes in an hour. He works for 7 hours. How many boxes does he pack in all?

Step 1: Read the problem.

Step 2: Learn what you must find out.

How many boxes did he pack in all? Clue Words: *in all, in an hour, for seven hours*

Step 3: Notice the clue words. Clue words help you solve the problem. In this problem, there are three groups of clue words. *In all* tells you to put numbers together. To do this you could add or multiply. But there are two other groups of clue words: *in an hour* and *for seven hours.*

These words suggest you should multiply.

Step 4: Copy down the numbers.

Numbers 16 7

Step 5: Write and solve the problem.

$$\begin{array}{r} 16 \\ \times 7 \\ \hline 112 \end{array}$$

Step 6: Check the answer. Ask yourself if it makes sense. For example, suppose you got an answer of 23. (16 + 7)

That answer would not make sense. If Norm can pack 16 boxes in one hour, he can surely pack more than 23 boxes in seven hours.

Clue Words

Words and groups of words like these give you a clue. They tell you to put numbers together. They tell you to multiply.

in five months *for* six hours *at* $13 each

Reminder: Not all problems have clue words.

PROBLEMS TO SOLVE

Solve the problems below on a separate sheet of paper.

1. Claudia saves $12 a week. How much does she save in a year? (Reminder: There are 52 weeks in a year.)
2. The United High School Chorus gave a concert. A total of 1,269 tickets were sold. Each ticket was sold for five dollars. How much money did the chorus make from ticket sales?
3. A program was sold at the concert. The program cost two dollars. A total of 649 programs were sold. How much money was made from selling programs?
4. John makes $8.00 an hour. If he works for 37 hours, how much money will he make?
5. Lydia works in a kennel. There are 37 puppies in the kennel. Each puppy eats 5 ounces of food a day. How many ounces of food do the puppies eat every day?
6. There are 72 envelopes in each package. How many envelopes are there in 569 packages?
7. Rebecca rented a car for 43 dollars a day. She kept the car for six days. How much did the car cost her?
8. There are 892 students at Center Art School. Each student pays $927 a year. In all, how much do the students pay?
9. Kim pays $68.00 a month for medical insurance. How much does she pay in a year? (Reminder: There are 12 months in a year.)
10. Nicole wanted to raise money for the hospital toy fund. She gave a dinner and charged 19 dollars a person. Twenty-eight people came to the dinner. How much money did Nicole raise?

4.11 Deciding How to Solve a Problem

You must read a problem carefully to decide how to solve it. Then you need to think about what a sensible answer might be. You must decide whether your answer should be larger or smaller than the largest number in the problem.

A *sensible* answer is reasonable. It makes *sense.*

If a sensible answer would be a larger number, you should add or multiply.

If a sensible answer would be a smaller number, you should subtract or divide.

MIXED PROBLEMS TO SOLVE

The problems below can be solved by adding, subtracting, or multiplying, Decide the best way to solve each problem. Then work them on a separate sheet of paper.

1. Doreen had $97. She bought a blouse for $34. How much money did she have left?
2. Mark spent $34 at the supermarket. He spent $18 at the cleaners and $14 at the shoe repair store. How much did he spend all together?
3. Marlene works five days a week. Each working day her expenses total $6. How much do her expenses come to each week?
4. Steve sells cars. He sold one car for $2,385. He sold another car for $1,967. He sold a third car for $3,548. How much did he get for all the cars?
5. Naomi worked at a temporary job for 17 weeks. She made $372 a week. How much did she make altogether?
6. Brett and Kurt went fishing. They caught 38 fish altogether. If Brett caught 17 fish, how many did Kurt catch?

MATHEMATICS IN YOUR LIFE: Figuring Your Pay

Connie has a part-time job at a grocery store. She works a different number of hours each day. She makes $6.00 an hour. Here are the hours Connie worked for the last two weeks.

Week Beginning 9/12		Week Beginning 9/19	
Day	**Hours Worked**	**Day**	**Hours Worked**
Monday	4	Monday	3
Tuesday	3	Tuesday	5
Wednesday	4	Wednesday	2
Thursday	6	Thursday	6
Friday	5	Friday	4

Answer the following questions on a separate sheet of paper.

1. How many hours did Connie work the week of 9/12?
2. How many hours did she work the week of 9/19?
3. Which week did she work longer? How many more hours did she work that week?
4. Connie makes $6.00 an hour. How much money did she make the week of 9/12?
5. How much did Connie make the week of 9/19?
6. How much less did Connie make the second week?
7. How much did Connie make for the two weeks altogether?
8. If Connie worked the same number of hours the next two weeks, how much would she make in a month?

4.12 Using Your Calculator: Multiplication

A calculator makes it easy to multiply large numbers. Multiply the example. Follow the steps below.

Step 1: Turn on your calculator. Find the [×] sign. Find the [=] sign.

Step 2: Press [9] [6] [7] [5]

Step 3: Press [×]

Step 4: Press [8] [3] [0] [4]

Step 5: Press [=]

Step 6: Your answer should be: 80,341,200

PRACTICE

A. Use a calculator to find the product of each pair of numbers. Write your answers on a separate sheet of paper.

1. 5,603 9,327 **3.** 92,843 4,729 **5.** 510, 999

2. 846, 4,962 **4.** 79,080 9,375 **6.** 753, 869

B. Use a calculator to find the product of each pair of numbers. Write your answers on a separate sheet of paper.

1. $2{,}388 \times 459$

2. $6{,}007 \times 468$

3. $6{,}853 \times 724$

4. $7{,}383 \times 2{,}875$

5. $6{,}020 \times 3{,}862$

6. $12{,}837 \times 6{,}355$

Chapter Review

CHAPTER SUMMARY

■ Multiply	To add a number to itself one or more times; to find the product of two numbers.	$5 \times 3 = 15$
■ Factors	Numbers being multiplied are called factors.	
■ Multiplying 0	When you multiply any number by 0, the answer is 0.	$5 \times 0 = 0$
■ Multiplying by 1	When you multiply any number by 1, the number stays the same.	$5 \times 1 = 5$
■ Multiplying Larger Numbers	To multiply larger numbers, start with the digits in the ones place. Then multiply one column at a time.	$\begin{array}{r} 232 \\ \times\ 32 \\ \hline 464 \\ 696 \\ \hline 7424 \end{array}$
■ Renaming	When the product of digits in a column is 10 or more, you must rename.	$\begin{array}{r} 697 \\ \times\ 4 \\ \hline 2788 \end{array}$
■ Estimating	Estimating gives you a quick, close idea of what an answer will be. Rounding numbers helps you estimate.	
■ Solving Problems	Decide whether your answer should be larger or smaller than the largest number in the problem. If the answer should be larger, you should add or multiply. If the answer should be smaller, you should subtract or divide.	

CHAPTER REVIEW

CHAPTER QUIZ

A. Copy the problems on a separate sheet of paper. Find the products.

1. 53×3

2. 700×69

3. $6{,}539 \times 2{,}935$

4. 632×21

5. 617×54

6. $7{,}843 \times 5{,}784$

7. 50×7

8. 724×73

9. $2{,}697 \times 752$

10. 927×0

11. 851×294

12. $26{,}207 \times 90$

B. Solve these word problems on a separate sheet of paper.

1. Bob still has 10 car payments to make. Each payment is $125. In all, how much does Bob still owe on his car?

2. There are 100 paper clips in each box. How many paper clips are there in 14 boxes?

3. Marty has 12 baskets of tomatoes. She needs 4 times that many to make spaghetti sauce. How many tomatoes does she need altogether?

4. Each bookshelf holds 27 books. How many books do 18 shelves hold?

5. Parker pays his mother $20.00 a week for room and board. In all, how much does Parker pay in a year (52 weeks)?

REVIEWING VOCABULARY

Rewrite the sentences below on a separate sheet of paper. Choose the correct word from the box to complete each sentence.

multiplication	product	partial product	factor

1. The ________ is not the final answer to a multiplication problem.

2. ________ is a quick way to add.

3. When you multiply two numbers together, you get a ________ .

4. A number being multiplied is called a ________ .

Dividing Whole Numbers

5

Each year 33 cars race in the famous Indy 500. They go around the track 200 times to complete the 500 mile championship race. In 1988 winning driver Rick Mears averaged 144 miles per hour. How many hours did it take Mears to complete the race? Find out by dividing that number into 500.

Chapter Learning Objectives

1. Divide whole numbers
2. Check multiplication and division
3. Find quotients with remainders
4. Divide larger numbers
5. Use zero in the quotient
6. Estimate answers to division problems
7. Use division to solve word problems

WORDS TO KNOW

- **divide** — to find out how many times one number contains another
- **dividend** — the number to be divided
- **divisor** — the number to divide by
- **quotient** — the number obtained by dividing one number into another; the answer in a division problem
- **remainder** — the number left over in a division problem
- **arithmetic operation** — addition, subtraction, multiplication, or division; each one is a separate arithmetic operation

5.1 What Is Division?

Division is the process of finding out how many times one number contains another.

Look at this division example. $15 \div 3 =$ This means $3\overline{)15}$.
This problem tells you to **divide** 15 by 3.

You could draw a picture like this.

quotient
divisor $\overline{)\text{dividend}}$
dividend ÷ divisor = quotient

Division is a quick way to subtract.

You could keep subtracting 3 from 15 until you got to 0.

$15 - 3 = 12$ $12 - 3 = 9$ $9 - 3 = 6$ $6 - 3 = 3$
$3 - 3 = 0$

To find the answer you had to subtract 5 times. Or you could solve the problem a faster way. You could divide.

PRACTICE

Write these division problems using numbers. The first one has been done for you.

1. Eight divided by two equals four $8 \div 2 = 4$

2. Six divided by three equals two

3. Twelve divided by four equals three

5.2 Basic Division

Remembering division facts makes dividing quicker and easier. One way to learn division is by studying the division chart below. It shows how to divide numbers through the number 36.

Suppose you want to divide 3 into 12. Follow the steps.

Step 1: Find the number 3 in the first column going down.

Step 2: Move across that row until you find the number 12.

Step 3: Move up that column until you come to the top row. The number in that box is your answer.

$3\overline{)12}$ with quotient 4 or $12 \div 3 = 4$

Study the chart below. Practice dividing the numbers in blue into the other numbers across each row.

1	2	3	4	5	6	7	8	9
2	4	6	8	10	12	14	16	18
3	6	9	12	15	18	21	24	27
4	8	12	16	20	24	28	32	36
5	10	15	20	25	30	35	40	45
6	12	18	24	30	36	42	48	54
7	14	21	28	35	42	49	56	63
8	16	24	32	40	48	56	64	72
9	18	27	36	45	54	63	72	81

Reading and Writing Division Problems

Division problems may be written two ways. Read the example:

15	÷	3	=	5	OR	$\begin{array}{r} 5 \\ 3\overline{)15} \end{array}$
Fifteen	divided by	three	equals	five.		

The result of a division problem is called the **quotient.**

PRACTICE

A. Read each problem below.

1. $12 \div 3 = 4$

2. $\begin{array}{r} 9 \\ 2\overline{)18} \end{array}$

3. $24 \div 8 = 3$

4. $\begin{array}{r} 9 \\ 7\overline{)63} \end{array}$

B. Copy the problems below on a separate sheet of paper. Find the quotients. If you need help, look at the division chart on page 349 in the back of the book.

1. $6 \div 2 =$

2. $6 \div 3 =$

3. $12 \div 3 =$

4. $12 \div 4 =$

5. $64 \div 8 =$

6. $16 \div 8 =$

7. $72 \div 8 =$

8. $27 \div 9 =$

9. $27 \div 3 =$

10. $81 \div 9 =$

11. $9 \div 3 =$

12. $30 \div 6 =$

13. $35 \div 5 =$

14. $35 \div 7 =$

15. $42 \div 6 =$

16. $16 \div 4 =$

17. $21 \div 7 =$

18. $32 \div 4 =$

19. $54 \div 6 =$

20. $48 \div 8 =$

C. Copy the problems below on a separate sheet of paper. Find the quotients. Try to solve the problems without using the division chart.

1. $7\overline{)56}$

2. $8\overline{)80}$

3. $6\overline{)48}$

4. $9\overline{)36}$

5. $4\overline{)32}$

6. $5\overline{)50}$

7. $3\overline{)18}$

8. $7\overline{)49}$

9. $2\overline{)18}$

10. $8\overline{)72}$

11. $6\overline{)36}$

12. $6\overline{)24}$

13. $9\overline{)81}$

14. $7\overline{)56}$

15. $4\overline{)40}$

16. $8\overline{)24}$

17. $4\overline{)20}$

18. $5\overline{)25}$

19. $7\overline{)42}$

20. $9\overline{)54}$

21. $6\overline{)54}$

22. $9\overline{)81}$

23. $5\overline{)45}$

24. $8\overline{)56}$

25. $7\overline{)49}$

Dividing a Number by Itself or One

When you divide any number by itself, the answer is 1.

$6 \div 6 = 1$ $234 \div 234 = 1$ $1{,}675 \div 1{,}675 = 1$

When you divide any number by one, the number stays the same.

$6 \div 1 = 6$ $234 \div 1 = 234$ $1{,}675 \div 1 = 1{,}675$

See for yourself. Look up $6 \div 6 = 1$ and $6 \div 1 = 6$ on the division chart.

PRACTICE

D. Copy the problems below on a separate sheet of paper. Find the quotients.

1. 1)65 **3.** 65)65 **5.** 4)4 **7.** 10)10

2. 8)8 **4.** 1)9 **6.** 1)16 **8.** 1)10

E. Copy the problems below on a separate sheet of paper. Find the quotients.

1. $8 \div 8 =$	**5.** $7 \div 1 =$	**9.** $32 \div 32 =$	**13.** $47 \div 1 =$	**17.** $72 \div 8 =$
2. $21 \div 3 =$	**6.** $14 \div 7 =$	**10.** $24 \div 8 =$	**14.** $14 \div 2 =$	**18.** $24 \div 6 =$
3. $32 \div 8 =$	**7.** $63 \div 7 =$	**11.** $50 \div 1 =$	**15.** $20 \div 4 =$	**19.** $16 \div 8 =$
4. $27 \div 9 =$	**8.** $72 \div 9 =$	**12.** $16 \div 4 =$	**16.** $56 \div 7 =$	**20.** $100 \div 100 =$

5.3 Dividing with More Than One Digit in the Answer

Solve this problem: 3)96 Follow the steps below.

Step 1: Ask yourself how many 3's there are in 9. Write 3 above the 9. 3

$$3\overline{)96}\text{ quotient }3$$

Step 2: Ask yourself how many 3's there are in 6. Write 2 above the 6. 2

$$3\overline{)96}\text{ quotient }32$$

PRACTICE

Copy the problems below on a separate sheet of paper.
Find the quotients.

1. $3\overline{)69}$
2. $3\overline{)39}$
3. $4\overline{)484}$
4. $4\overline{)88}$
5. $2\overline{)82}$
6. $2\overline{)648}$
7. $2\overline{)64}$
8. $5\overline{)55}$
9. $3\overline{)936}$
10. $6\overline{)66}$
11. $4\overline{)48}$
12. $2\overline{)2462}$

5.4 Checking Multiplication and Division

It isn't hard to check the answer to a division problem. Just multiply the quotient by the divisor.

The number you divide by in a division problem is called the *divisor.*

$$\begin{array}{r} 2,143 \\ 2\overline{)4,286} \end{array} \qquad \begin{array}{r} 2,143 \\ \times \quad 2 \\ \hline 4,286 \end{array}$$

The answer to a multiplication problem can be checked by dividing the product by one of the factors.

Remember: The answer to a multiplication problem is the *product.* The numbers to be multiplied are called *factors.*

$$\begin{array}{r} 2,143 \\ \times \quad 2 \\ \hline 4,286 \end{array} \qquad \begin{array}{r} 2,143 \\ 2\overline{)4,286} \end{array}$$

PRACTICE

Copy each problem below on a separate sheet of paper.
Find the quotients. Check your answers by multiplying.

1. $2\overline{)862}$
2. $5\overline{)555}$
3. $4\overline{)448}$
4. $2\overline{)268}$
5. $3\overline{)639}$
6. $2\overline{)488}$
7. $4\overline{)884}$
8. $3\overline{)993}$

5.5 Dividing with Remainders

All problems don't divide evenly. Sometimes a number is left over. This number is called a **remainder.** The remainder is part of the quotient.

You can find these quotients easily: $\begin{array}{r}2\\2\overline{)\,4}\end{array}$ $\begin{array}{r}3\\2\overline{)\,6}\end{array}$

But this problem leaves a remainder: $\begin{array}{r}2 \text{ R } 1\\2\overline{)\,5}\phantom{\text{ R } 1}\end{array}$

It isn't hard to solve problems that leave remainders. Follow the steps below to solve this problem: $3\overline{)97}$

Step 1: Ask yourself how many 3's there are in 9.

Write 3 above the 9.

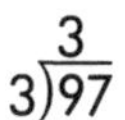

Step 2: Ask yourself how many 3's there are in 7.

$\begin{array}{r}32 \text{ R } 1\\3\overline{)97}\phantom{\text{ R } 1}\end{array}$

You know that there are at least two because $3 \times 2 = 6$.

But there can't be three because $3 \times 3 = 9$.

So, the answer is 2. Write 2 above the 7. Then write R 1 to show that there is 1 left over. The remainder is 1.

PRACTICE

Copy each problem below on a separate sheet of paper. Find the quotients.

1. $3\overline{)62}$

2. $4\overline{)86}$

3. $2\overline{)49}$

4. $5\overline{)257}$

5. $3\overline{)638}$

6. $5\overline{)557}$

7. $2\overline{)499}$

8. $4\overline{)487}$

9. $3\overline{)365}$

5.6 Dividing Larger Numbers

It is easy to divide larger numbers if you follow certain steps.

Solve this problem: $6\overline{)583}$ Follow the steps below.

Step 1: Ask yourself how many 6's there are in 5. Six does not divide into 5. Do not write anything above the 5.

$$6\overline{)583}$$

Step 2: Since you cannot work with 5 alone, work with 58. Notice that 6 cannot be divided into 58 without a remainder. So think of numbers smaller than 58. What is the closest number to 58 that 6 can divide into evenly? The answer is 54. $54 \div 6 = 9$

Write 9 above the 8.

$9 \times 6 = 54$

Write 54 under 58.

$58 - 54 = 4$

$$\begin{array}{r} 9 \\ 6\overline{)583} \\ \underline{54} \\ 4 \end{array}$$

Step 3: Bring down the 3.

$$\begin{array}{r} 9 \\ 6\overline{)583} \\ \underline{54} \\ 43 \end{array}$$

Step 4: Ask yourself how many 6's there are in 43.

$42 \div 6 = 7$

There are 7 with a remainder of 1.

Write 7 above the 3.

Write 42 under 43.

$43 - 42 = 1$ Write R 1

$$\begin{array}{r} 97 \text{ R } 1 \\ 6\overline{)583}\phantom{\text{ R } 1} \\ \underline{54}\phantom{\text{ R } 1} \\ 43\phantom{\text{ R } 1} \\ \underline{42}\phantom{\text{ R } 1} \\ 1\phantom{\text{ R } 1} \end{array}$$

PRACTICE

Copy the problems below on a separate sheet of paper. Find the quotients.

1. $6\overline{)497}$

2. $2\overline{)57}$

3. $4\overline{)653}$

4. $5\overline{)320}$

5. $9\overline{)704}$

6. $8\overline{)697}$

7. $6\overline{)382}$

8. $5\overline{)276}$

9. $3\overline{)795}$

10. $7\overline{)589}$

11. $2\overline{)512}$

12. $7\overline{)819}$

13. $4\overline{)536}$

14. $6\overline{)729}$

15. $8\overline{)721}$

16. $4\overline{)339}$

17. $8\overline{)953}$

18. $3\overline{)298}$

19. $6\overline{)493}$

20. $9\overline{)323}$

5.7 Dividing by Numbers with More Than One Digit

You can divide by numbers with more than one digit by following certain steps.

Solve this problem: $26\overline{)591}$ Follow the steps below.

Step 1: You are dividing by a two-digit number. Since 5 is smaller than 26, begin working with 59. Ask yourself how many 26's there are in 59. Notice that 26 cannot be divided into 59 without a remainder.

$$\begin{array}{r} 2 \\ 26\overline{)591} \\ \underline{52} \\ 7 \end{array}$$

So think of numbers smaller than 59. What is the closest number to 59 that 26 can divide into without a remainder? The answer is 52.

$$\begin{array}{r} 22 \text{ R } 19 \\ 26\overline{)591} \\ \underline{52} \\ 71 \\ \underline{52} \\ 19 \end{array}$$

$52 \div 26 = 2$

Write 2 above the 9.

$2 \times 26 = 52$

Write 52 under 59.

$59 - 52 = 7$

Step 2: Bring down the 1.

Step 3: How many 26's are there in 71?

$71 \div 26 = 2$ with a remainder.

$71 - 52 = 19$

Write 2 above the 1. Write R 19.

PRACTICE

Copy the problems below on a separate sheet of paper. Find the quotients. Be sure to write the remainder when there is one.

1. $32\overline{)585}$
2. $41\overline{)870}$
3. $92\overline{)976}$
4. $15\overline{)973}$
5. $12\overline{)145}$
6. $17\overline{)829}$
7. $23\overline{)852}$
8. $16\overline{)219}$
9. $37\overline{)748}$
10. $32\overline{)992}$
11. $25\overline{)347}$
12. $41\overline{)693}$
13. $15\overline{)358}$
14. $20\overline{)480}$
15. $55\overline{)605}$
16. $64\overline{)870}$
17. $52\overline{)537}$
18. $67\overline{)803}$
19. $40\overline{)487}$
20. $25\overline{)383}$

5.8 Using Zeros in the Answer

You have already learned the steps in solving division problems.

Step 1: Divide the first part of the dividend by the divisor. Write the digit above the dividend.

Step 2: Multiply that digit by the divisor.

Step 3: Subtract.

Step 4: Bring down the next digit.

Step 5: Repeat these steps until each digit in the dividend has been used.

Sometimes you need to put a zero in an answer.

Study this example:

The 3 has been brought down.
Since 6 cannot be divided into 3, write 0
above the 3 in the quotient.

$$\begin{array}{r} 80 \\ 6\overline{)4,834} \\ \underline{4\,8}\downarrow \\ 03 \end{array}$$

Bring the 4 from the dividend down.
Now you can divide 6 into 34.
Write 5 in the quotient.
Multiply: $5 \times 6 = 30$.
Then subtract 30 from 34.
Can 6 be divided into 4? No.
Are there any more digits to bring down?
No.
So the remainder is 4.

$$\begin{array}{r} 805 \text{ R } 4 \\ 6\overline{)4,834}\phantom{\text{ R } 4} \\ \underline{4\,8}\downarrow\phantom{\text{ R } 4} \\ 034\phantom{\text{ R } 4} \\ \underline{30}\phantom{\text{ R } 4} \\ 4\phantom{\text{ R } 4} \end{array}$$

PRACTICE

Copy the problems below on a separate sheet of paper.
Find the quotients.

1. $8\overline{)6,472}$
2. $24\overline{)4,832}$
3. $35\overline{)7,080}$
4. $2\overline{)1,213}$
5. $18\overline{)1,725}$
6. $26\overline{)2,493}$
7. $5\overline{)3,029}$
8. $11\overline{)6,653}$
9. $50\overline{)5,032}$
10. $4\overline{)1,695}$
11. $27\overline{)5,492}$
12. $14\overline{)4,227}$

5.9 Estimating and Thinking

You have learned that rounding numbers can help you estimate what an answer will be. It also helps to ask yourself if an estimated answer seems sensible. Think it through. An answer that "makes sense" is neither too big nor too small to be possible.

Read the example below.

You are bringing hamburgers to a barbecue. There will be a total of 20 people there. You think that most people will eat two hamburgers. How many hamburgers should you bring? 20 75 40

20 Think about it. This answer is too small to make sense. Nobody can have two hamburgers if there are only 20 hamburgers for 20 people.

75 This answer is too big to make sense. Everybody would have three hamburgers and there would still be some left over.

40 This is the exact answer. Remember, though, that a sensible estimate does not have to be the exact answer. It just has to be close.

PRACTICE

Read the problems below. Choose the estimates that are closest to the exact answers in each problem. Write your answers on a separate sheet of paper. Be prepared to explain your answers.

1. Shelley drove 354 miles. She used 20 gallons of gas. Estimate about how many miles she traveled on each gallon.

33 6 18

2. The round-trip distance between Union City and Jefferson is 69 miles. Marsha made the trip 7 times last month. Estimate about how many miles she drove in all.

500 1,200 250

3. Eight people can sit at each table in the high school cafeteria. About 500 students have lunch during each lunch period. Estimate about how many tables are needed.

250 65 100

5.10 Using Division to Solve Word Problems

Steps in Solving Word Problems.

Read this problem. Solve it by following the steps below.

Marilyn sold 17 sets of glasses. Her sales totaled $374. How much did each set cost?

Step 1: Read the problem.

Step 2: Learn what you must find out.

How much did each set cost?

Step 3: Notice the clue words. Clue words help you solve the problem. In this problem, there is one group of clue words. *How much did each* tells you that you need to divide.

Clue Words: *how much did each*

Step 4: Copy down the numbers.

Numbers: 17 $374

Step 5: Write and solve the problem.

$$\begin{array}{r} \$22 \\ 17\overline{)\$374} \\ \underline{34} \\ 34 \end{array}$$

Step 6: Check the answer. Ask yourself if it makes sense.

Clue Words

Words and groups of words like these give you a clue. They tell you to separate numbers into groups. They tell you to divide.

how much did each *how many times* *into how many*

Reminder: Not all problems have clue words.

PROBLEMS TO SOLVE

Solve these problems on a separate sheet of paper.

1. The symphony sold $852 worth of tickets to its opening concert. A total of 71 people attended the concert. How much did each ticket cost?
2. On Stan's last sales trip, he drove 2,763 miles. The trip lasted 14 days. How many miles did Stan drive each day?
3. Dan worked for 23 days. He made $1,325. How much did he make each day?
4. Dave and Bev sold the same number of tickets to the dance. Between them they sold 56. How many did each sell?
5. Russell worked for 4 days last week. He made 72 phone calls. How many calls did he make a day?

MIXED PROBLEMS

Addition, subtraction, multiplication, and division are called **arithmetic operations.**

Read these problems carefully. Decide which operation you need to use to solve each one. Then solve the problems on a separate sheet of paper.

1. Catherine designs and makes leather handbags. It takes her three days to make a handbag. Last year she made 88 handbags. How many days did she work?
2. Jack bought a five-pound bag of flour. He also bought two pounds of potatoes and three pounds of apples. How many pounds of groceries did he have to carry?
3. Hank owns 67 records. Penny has 49. How many more records does Hank have?
4. Sid collects stamps. He has 3,453 from the United States. He has 679 from Argentina. He has 856 from Italy. How many stamps does he have altogether?

MATHEMATICS IN YOUR LIFE: Medication Labels

Karen was using two kinds of medication for her cold and cough. Each had a label on the bottle. The label told how much medication could be taken safely. It also told how often the medication could be taken safely. Read the labels.

Medication A—Cold Capsules

Dosage
Adults: 2 capsules every 4 hours
Do not take more than 8 capsules in 24 hours.
Children: 1 capsule every 4 hours
No more than 4 capsules in 24 hours.

Medication B—Cough Medicine

Dosage
Adults: 2 teaspoons every 6 hours
Do not take more than 6 teaspoons in 24 hours.
Children: 1 teaspoon every 6 hours.
No more than 3 teaspoons in 24 hours.

1. Karen is an adult. How many capsules of medication A can she take in 24 hours?
2. How many capsules of medication A can she take at one time?
3. How long should she wait between taking those capsules?
4. How many capsules of medication A can be given to a child in 24 hours?
5. How long should a child wait between taking capsules of medication A?
6. How many teaspoons of medication B should Karen take at one time?
7. Which medication can an adult take more often in 24 hours?

5.11 Using Your Calculator: Division

You have learned how useful a calculator is to solve problems. But you can also have fun with your calculator. Amaze your friends with this trick.

Step 1: Ask a friend to enter any three digits on a calculator.

Step 2: Ask your friend to enter the same digits in the same order again. This will make a six-digit number.

Step 3: Tell your friend that the number can be divided by 13 without a remainder. Have your friend divide by 13.

Step 4: Then tell your friend that the quotient that resulted can be divided by 11 without a remainder. Have your friend divide by 11.

Step 5: Next tell your friend to divide by 7 to get the original three-digit number.

Before trying this trick on anyone else, do it yourself.

PRACTICE

Try these examples to see if the trick works.

1. 478

2. 523

3. 657

4. 921

5. 368

Chapter Review

CHAPTER SUMMARY

■ Divide	To divide means to find out how many times one number contains another.
■ Dividing a number by itself	When you divide any number by itself, the answer is 1. $5 \div 5 = 1$
■ Dividing a number by 1	When you divide any number by 1, the number stays the same. $5 \div 1 = 5$
■ Checking Multiplication and Division	You can check the answer to a division problem by multiplying the quotient by the divisor. You can check the answer to a multiplication problem by dividing the product by either of the factors.
■ Remainder	Sometimes a number does not divide evenly into another. Then the quotient has a remainder.

REVIEWING VOCABULARY

Copy the sentences below on a separate sheet of paper. In each sentence use the correct word or group of words from the box.

divide	quotient	remainder	arithmetic operation	divisor	dividend

1. The ________ is the number obtained by dividing one number into another.

2. Multiplication is an ________ .

3. The number you divide by in a division problem is called the ________ .

4. You ________ to find out how many times one number contains another number.

5. The ________ is the number left over in a division problem.

6. The number to be divided in a division problem is called the ________ .

Chapter Quiz

A. Copy the problems below on a separate sheet of paper. Find the quotients. Write the remainder when there is one.

1. $1\overline{)37}$

2. $12\overline{)\ 12}$

3. $35\overline{)\ 76}$

4. $52\overline{)164}$

5. $4\overline{)8,792}$

6. $31\overline{)\ 350}$

7. $87\overline{)\ 935}$

8. $62\overline{)\ 634}$

9. $25\overline{)250}$

10. $47\overline{)843}$

11. $50\overline{)\ 55}$

12. $83\overline{)896}$

13. $21\overline{)\ 859}$

14. $33\overline{)\ 990}$

15. $17\overline{)6,750}$

16. $12\overline{)2,963}$

17. $43\overline{)\ 562}$

18. $59\overline{)6,182}$

19. $29\overline{)3,132}$

20. $37\overline{)\ 925}$

B. Developing Thinking Skills

Read the problem below. Then answer the questions on a separate sheet of paper.

Earl was supposed to work for 4 hours on Saturday. At the end of his shift he was supposed to pack 24 crates of oranges. Each crate had 48 oranges in it. How many total oranges would Earl have to pack? How many oranges would Earl have to pack each hour to get the job done?

1. What are you asked to find out?

2. What are the clue words?

3. What do the clue words tell you to do?

4. What are the numbers?

5. What are the answers to the problem?

C. Solve these problems on a separate sheet of paper.

1. Phyllis works 40 hours each week. If she works the same number of hours each day and she works 5 days each week, how many hours does she work each day?

2. Tina and Barbara sold the same number of raffle tickets. Together they sold 108. How many did each girl sell?

MORE ABOUT NUMBERS

6

In the early 20th century millions of immigrants found a new home in America. Between 1890 and 1920 the U.S. population increased by 43 million people. More than half this number—18 million—were immigrants. Upon landing, each of the immigrants was given a physical examination, including an eye check, as shown here.

Chapter Learning Objectives

1. Write the rules of divisibility
2. Find factors
3. Find Greatest Common Factors
4. Find multiples
5. Find Least Common Multiples
6. Find prime and composite numbers
7. Find prime factors

WORDS TO KNOW

■ **divisible**	can be divided evenly without a remainder
■ **greatest common factor**	the largest factor that two or more products share; remember that a factor is one of the numbers that must be multiplied to obtain a product
■ **multiples of a number**	the products of multiplying that number by whole numbers
■ **least common multiple**	the smallest multiple that two or more numbers share
■ **prime number**	a number with only two factors—itself and 1
■ **composite number**	a number with more than two factors
■ **infinite**	without end

6.1 Divisibility Tests

A number that can be divided by a second number evenly without a remainder is *divisible* by the second number.

Remember: a number is even if it ends in 0,2,4,6, or 8.

18 is divisible by 3.

$18 \div 3 = 6$

16 is divisible by 4.

$16 \div 4 = 4$

15 is divisible by 5.

$15 \div 5 = 3$

24 is divisible by 8.

$24 \div 8 = 3$

You can sometimes tell if a number is divisible by another number by using the divisibility tests.

Divisibility Tests

Divisible* by *2 A number is divisible by 2 if it is an even number.

$126 \div 2 = 63$

Divisible* by *4 A number is divisible by 4 if the last two digits are divisible by 4. $628 \div 4$

$28 \div 4 = 7$

So 628 is divisible by 4; $628 \div 4 = 157$

Divisible* by *6 A number is divisible by 6 if it can be divided by both 2 and 3. $30 \div 6$

$30 \div 2 = 15$ $30 \div 3 = 10$

So 30 is divisible by 6; $30 \div 6 = 5$

Divisible* by *10 A number is divisible by 10 if the last digit is zero. $80 \div 10 = 8$

Divisible* by *3 A number is divisible by 3 if the sum of the digits can be divided by 3. $471 \div 3$

$4 + 7 + 1 = 12$; $12 \div 3 = 4$

So 471 is divisible by 3; $471 \div 3 = 157$

Divisible* by *5 A number is divisible by 5 if the last digit is 0 or 5.

$6{,}840 \div 5 = 1{,}368$

Divisible* by *9 A number is divisible by 9 if the sum of the digits can be divided by 9. $423 \div 9$

$4 + 2 + 3 = 9$; $9 \div 9 = 1$

So 423 is divisible by 9; $423 \div 9 = 47$

PRACTICE

Use the divisibility tests to answer these questions. Write your answers on a separate sheet of paper. The first one has been done for you.

1. Is 3,564 divisible by 2?

   Yes. (It is an even number.)
2. Is 8,916 divisible by 4?
3. Is 3,680 divisible by 10?
4. Is 5,334 divisible by 3?
5. Is 9,504 divisible by 6?
6. Is 7,097 divisible by 9?
7. Is 4,727 divisible by 4?
8. Is 3,881 divisible by 2?
9. Is 9,566 divisible by 6?
10. Is 2,615 divisible by 5?

6.2 Factors

The numbers you multiply to obtain a product are called factors of that product.

factor × factor = product
F × F = P

Here are the factors of 12:

1 × 12
2 × 6
3 × 4

The factors of 12 are: 1,2,3,4,6, and 12.
The factors of 12 can be written this way:
$F_{12} = \{1,2,3,4,6,12\}$.

Did You Know?
An adult heart beats an average of 100,000 times a day. If a person reaches the age of 75, his or her heart will have beaten $2\frac{1}{2}$ billion times.

You can find all of the factors of a number by trying the whole numbers in order. That means you *divide* the given number by the numbers 1,2,3, and so on. If the numbers divide exactly, then the divisor and the quotient are both factors. Keep dividing by the whole numbers until the factors repeat. For example: find the factors of 20.

Remember: the divisor in a division problem is the number you *divide by*. The answer you get is called the *quotient*.

F_{20}

divisor - $1\overline{)20}$, quotient - 20	1 divides 20 exactly, so 1 and 20 are factors. $1 \times 20 = 20$
$2\overline{)20}$ = 10	2 divides 20 exactly, so 2 and 10 are factors. $2 \times 10 = 20$
$3\overline{)20}$ = 6 R 2	3 does not divide 20 exactly, so there are no factors here.
$4\overline{)20}$ = 5	4 divides 20 exactly, so 4 and 5 are factors. $4 \times 5 = 20$
$5\overline{)20}$ = 4	5 divides 20 exactly, but the numbers repeat. Go no further.

The factors of 20 are: 1,2,4,5,10, and 20. Or you can write it this way: $F_{20} = \{1,2,4,5,10,20\}$.

PRACTICE

Find the factors of each number below. Write your answers on a separate sheet of paper.

1. F_{10}
2. F_{16}
3. F_{27}
4. F_{42}
5. F_{20}
6. F_{7}
7. F_{28}
8. F_{30}
9. F_{5}
10. F_{35}
11. F_{25}
12. F_{14}
13. F_{12}
14. F_{8}
15. F_{36}
16. F_{50}

6.3 Multiples

The multiples of a number are the products you get when you multiply that number by whole numbers.

M stands for multiple.

The numbers in color are some of the multiples of 8.

$8 \times 0 = 0$ $8 \times 1 = 8$ $8 \times 2 = 16$

$8 \times 3 = 24$ $8 \times 4 = 32$ $8 \times 5 = 40$

And so on.

We can write this as $M_8 = \{0, 8, 16, 24, 32, 40, \ldots\}$
The three dots show that the multiples continue on. They do not end. They are *infinite.*

PRACTICE

Find five multiples of each number below. Write your answers on a separate sheet of paper.

1. M_2 **2.** M_3 **3.** M_7 **4.** M_9 **5.** M_{12}

6.4 Prime and Composite Numbers

A prime number has only two factors. These factors are the number itself and 1.

The number 3 is a prime number. Its only factors are 3 and 1.

A composite number has more than two factors.

The number 6 is a composite number. Its factors are 1,2,3, and 6.

The number one is neither a prime number nor a composite number.

It has only one factor, the number 1.

PRACTICE

Find the factors of each number below. If it is a prime number, write *P* after the factors. If it is a composite number, write *C*. The first one has been done for you.

1. F_6 $\;1\overline{)6}^{\,6}$ $\;1 \times 6 = 6$ $\;F_6 = 1,2,3,6;$ C.

$2\overline{)6}^{\,3}$ $\;2 \times 3 = 6$

$3\overline{)6}^{\,2}$ Factors repeat

2. F_{49}

3. F_{54}

4. F_9

5. F_{15}

6. F_{60}

7. F_{32}

8. F_{12}

9. F_{48}

10. F_{24}

11. F_{40}

6.5 Finding Prime Numbers

Eratosthenes was a Greek mathematician who lived in the third century, B.C. He invented a way to find prime numbers.

His invention came to be known as Eratosthenes's sieve.

A real *sieve* is a strainer used to drain away water and other liquids. You can use Eratosthenes's sieve to drain away composite numbers.

Follow the instructions below to learn how to find all the prime numbers between 1 and 100.

Copy the chart on the following page. Numbers that are not prime numbers should be crossed out. Follow the steps.

1	2	3	4	5	6	7	8	9	10
11	12	13	14	15	16	17	18	19	20
21	22	23	24	25	26	27	28	29	30
31	32	33	34	35	36	37	38	39	40
41	42	43	44	45	46	47	48	49	50
51	52	53	54	55	56	57	58	59	60
61	62	63	64	65	66	67	68	69	70
71	72	73	74	75	76	77	78	79	80
81	82	83	84	85	86	87	88	89	90
91	92	93	94	95	96	97	98	99	100

Step 1: Cross out 1. It's not a prime number.

Step 2: The first prime number is 2. Circle it. Then cross out every multiple of 2. HINT: Every *even* number is a multiple of 2.

Step 3: The next prime number is 3. Circle it. Then cross out every multiple of 3, such as 6,9,12 . . .

Step 4: Repeat these steps. Every time you find a prime number, circle it. Then cross out all its multiples.

The numbers that are not crossed out when you are finished are the prime numbers less than 100. There are twenty-five of them.

6.6 Factoring to Prime Numbers

Every composite number can be shown as the product of prime numbers. Doing this is called prime factorization.

Finding the prime factors of a composite number is not difficult. The easiest way is to make a factor tree. For most numbers you can make several different factor trees. Follow the steps below.

Step 1: Write the number.

Step 2: Choose two numbers whose product is this number. These numbers do not have to be prime numbers.

Step 3: Find the factors of the numbers chosen in step 2. Keep repeating this step until all the factors are prime numbers. Sometimes you will get a prime number on a line. This happens in example C. Just bring that prime number down to the line below.

Step 4: Write the prime numbers.

Examples:

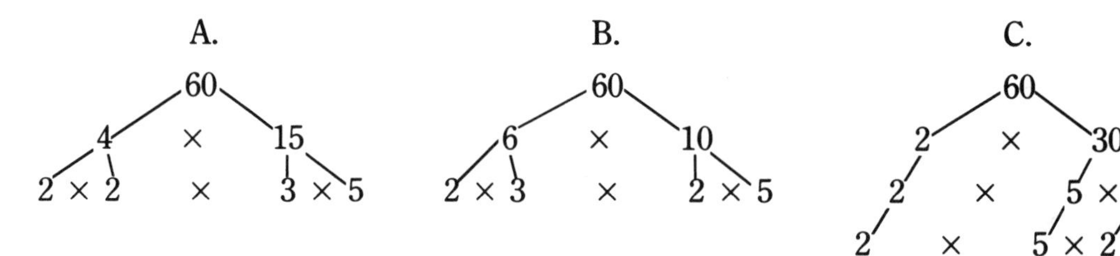

The prime factors of 60 are 2, 3, and 5.

PRACTICE

Find the prime factors of each number below by making factor trees. Do your work on a separate sheet of paper.

1. 24
2. 27
3. 48
4. 100
5. 32
6. 17
7. 70
8. 15
9. 64
10. 54
11. 32
12. 20
13. 50
14. 28
15. 56
16. 82
17. 40
18. 16
19. 12
20. 26

6.7 Least Common Multiple

You have already learned how to find the multiples of numbers. For example, the multiples of 6 are found this way: $6 \times 0 = 0$; $6 \times 1 = 6$; $6 \times 2 = 12$; $6 \times 3 = 18$, and so on. The numbers in blue are multiples of 6.

The multiples of a number are the products you get when you multiply that number by whole numbers.

It is also possible to compare the multiples of numbers.

Look at the multiples of these numbers.

$M_3 = (0, 3, 6, 9, 12, 15, 18, 21, 24, 27, 30, 33, 36, \ldots)$
$M_4 = (0, 4, 8, 12, 16, 20, 24, 28, 32, 36, 40, \ldots)$

Zero is the first multiple of every number. The product of any number and 0 is 0.

The example above shows some other multiples that 3 and 4 have in common. They are 12, 24, and 36.

The smallest multiple 3 and 4 have in common is 12. This multiple is called the *Least Common Multiple* or *LCM*.

It is easy to find the least common multiple of two or more numbers. Follow the steps below.

Step 1: Find the first few multiples of each number.

Step 2: Compare the lists of multiples. Find the lowest multiple, except 0, that is in all the lists.

Step 3: If you cannot find a common multiple, list more multiples.

PRACTICE

A. Copy each pair of numbers below on a separate sheet of paper. Next to each pair, write their Least Common Multiple.

1. 15, 20
2. 7, 3
3. 2, 5
4. 6, 8
5. 4, 12
6. 3, 30
7. 3, 9
8. 5, 6
9. 7, 8
10. 4, 5
11. 12, 16
12. 4, 6

B. Write the Least Common Multiple of these groups of numbers on a separate sheet of paper.

1. 2, 3, 5
2. 3, 7, 9
3. 3, 4, 6
4. 4, 7, 14
5. 5, 10, 20
6. 5, 6, 12
7. 4, 8, 2
8. 2, 7, 3

6.8 Greatest Common Factor

It is possible to compare the factors of two or more numbers.

Look at the factors of 24 and 36.

24		36	
$\begin{array}{r} 24 \\ 1\overline{)24} \end{array}$	$1 \times 24 = 24$	$\begin{array}{r} 36 \\ 1\overline{)36} \end{array}$	$1 \times 36 = 36$
$\begin{array}{r} 12 \\ 2\overline{)24} \end{array}$	$2 \times 12 = 24$	$\begin{array}{r} 18 \\ 2\overline{)36} \end{array}$	$2 \times 18 = 36$
$\begin{array}{r} 8 \\ 3\overline{)24} \end{array}$	$3 \times 8 = 24$	$\begin{array}{r} 12 \\ 3\overline{)36} \end{array}$	$3 \times 12 = 36$

$4\overline{)24}$ = 6 $4 \times 6 = 24$ $4\overline{)36}$ = 9 $4 \times 9 = 36$

$5\overline{)24}$ = 4 R 4   *No factors* $5\overline{)36}$ = 7 R 1   *No factors*

$6\overline{)24}$ = 4   *Factors repeat* $6\overline{)36}$ = 6   *Factors repeat*

$F_{24} = (1,2,3,4,6,8,12,24)$ $F_{36} = (1,2,3,4,6,9,12,18,36)$

The numbers 24 and 36 have these factors in common: 1, 2, 3, 4, 6, 12.

The number 12 is the largest of the common factors of 24 and 36. Therefore, 12 is called the *Greatest Common Factor,* or *GCF,* of 24 and 36. It is easy to find the greatest common factor of two or more numbers. Follow the steps below.

Step 1: Write the numbers whose greatest common factor you need to find.

Step 2: Divide the given numbers by whole numbers to find their factors. List the factors of each number separately.

Step 3: Find the largest shared factor. This is the Greatest Common Factor.

PRACTICE

Find the Greatest Common Factor of these pairs of numbers. Do your work on a separate sheet of paper.

1. 8, 15
2. 42, 60
3. 15, 20
4. 8, 12
5. 6, 9
6. 18, 15
7. 2, 5
8. 3, 10
9. 12, 24
10. 4, 18
11. 6, 15
12. 14, 49
13. 9, 30
14. 8, 20
15. 12, 15
16. 5, 11

MATHEMATICS IN YOUR LIFE:
Reading an Electric Meter

Power companies find out how much electricity you use by reading your electric meter. An electric meter keeps a running total of how many kilowatt-hours of electricity have been used. The electric company usually sends someone out to read the meter each month.

An electric meter has five dials. The dials are read from left to right.

This meter shows 23,548 as the current number of kilowatt-hours. How can you tell how many kilowatt-hours were used during the month? You would have to see what numbers this meter showed 30 days ago. Then you would subtract last month's numbers from 23,548. The difference would be the number of kilowatt-hours used during the month.

Read the numbers on each of the meters below. Then solve the word problems on a separate sheet of paper.

August 1

1. The meter above was read on August 1. One month before, on July 1, the same meter showed 10,455 kilowatt hours. How many kilowatt-hours were used during the month of July?

September 1

2. The meter above is the same one shown in problem #1. The above reading was made on September 1. How many kilowatt hours were used during the month of August?
3. During the month of September the customer used 2,422 kilowatt-hours. What numbers will the meter show when it is read on October 1?

6.9 Using Your Calculator: Factors

A calculator can make it easier to find the factors of larger numbers. Begin with 1 and try the whole numbers in order. Divide the whole numbers into the number whose factors you want to find. Using a calculator can help you do this quickly and correctly.

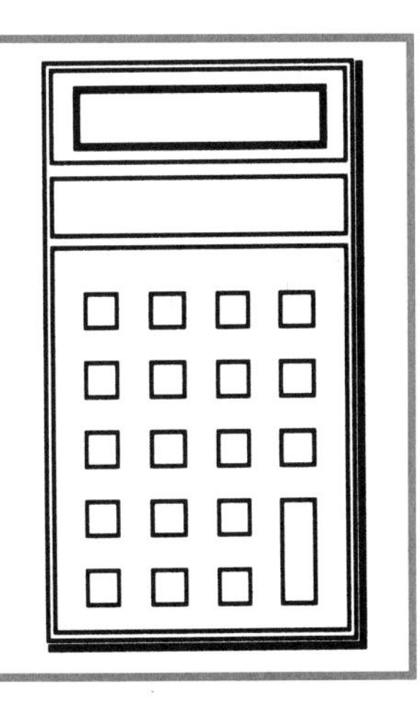

PRACTICE

Use a calculator to find the factors of these numbers.

1. 96
2. 150
3. 75
4. 136
5. 88
6. 66
7. 100
8. 108

CHAPTER REVIEW

CHAPTER SUMMARY

■ **Divisible**	When a number can be divided by a second number evenly, that number is divisible by the second number. Divisibility tests show whether a number is divisible by another number.
■ **Factors**	The numbers multiplied to obtain a product are the factors of that product. $F_4 = \{1,2,4\}$
■ **Greatest Common Factor**	The largest factor that two or more numbers share is their Greatest Common Factor (GCF). The GCF of 4 and 16 is 4.
■ **Multiples**	Multiples are the products obtained by multiplying a number by whole numbers. The multiples of any whole number are infinite. $M_4 = \{0,4,8,\ldots\}$
■ **Least Common Multiple**	The smallest multiple that two or more numbers share is their least common multiple (LCM). The LCM of 4 and 16 is 16.
■ **Prime Number**	A prime number has only two factors, itself and 1. Prime numbers: 2,3,5,. . .
■ **Composite Number**	A composite number has more than two factors. Composite numbers: 4,6,8,9,. . .

REVIEWING VOCABULARY

Number a separate sheet of paper from 1 to 4. Read each word in the column on the left. Find its definition in the column on the right. Write the letter of the definition next to the correct number.

1. infinite	a. can be divided evenly without a remainder
2. greatest common factor	b. one of the numbers that must be multiplied to obtain a product
3. divisible	c. the largest factor that two or more numbers share
4. factor	d. without end

CHAPTER REVIEW

CHAPTER QUIZ

A. Copy the chart below on a separate sheet of paper. Use divisibility tests to decide whether the numbers are divisible by 3,4,6, and 9. Then write either Yes or No in the appropriate column.

Number	Divisible by 3	Divisible by 4	Divisible by 6	Divisible by 9
612				
837				
1,215				
3,475				
4,444				
6,258				
9,128				
11,472				
15,650				
18,752				

B. Find the Greatest Common Factor of each pair of numbers below. Do your work on a separate sheet of paper.

1. 48, 20
2. 6, 24
3. 8, 18
4. 4, 44
5. 24, 30
6. 54, 9
7. 6, 32
8. 8, 48

C. Find the Least Common Multiple of each pair of numbers below. Do your work on a separate sheet of paper.

1. 3, 8
2. 7, 9
3. 4, 20
4. 11, 6
5. 5, 7
6. 12, 15
7. 24, 9
8. 22, 3

UNIT ONE REVIEW

Solve these problems on a separate sheet of paper.

1. $\begin{array}{r} 253 \\ 445 \\ +301 \\ \hline \end{array}$

2. $\begin{array}{r} 4,208 \\ 7,530 \\ +6,452 \\ \hline \end{array}$

3. $\begin{array}{r} 869 \\ -562 \\ \hline \end{array}$

4. $\begin{array}{r} 953 \\ -267 \\ \hline \end{array}$

5. $\begin{array}{r} 3,462 \\ -1,593 \\ \hline \end{array}$

6. $\begin{array}{r} 4,805 \\ -2,937 \\ \hline \end{array}$

7. $\begin{array}{r} 426 \\ \times\ 23 \\ \hline \end{array}$

8. $\begin{array}{r} 792 \\ \times 307 \\ \hline \end{array}$

9. $4\overline{)3,684}$

10. $2\overline{)869}$

11. $45\overline{)2,796}$

12. $37\overline{)5,609}$

13. Find the greatest common factor of each pair of numbers.
a. 16, 24 **b.** 21, 18 **c.** 50, 100 **d.** 20, 16

14. Find the least common multiple of each pair of numbers.
a. 3, 9 **b.** 24, 16 **c.** 55, 80 **d.** 6, 18

15. Write only the prime numbers.

1	2	3	4	5	6	7	8	9	10
11	12	13	14	15	16	17	18	19	20
21	22	23	24	25	26	27	28	29	30
31	32	33	34	35	36	37	38	39	40
41	42	43	44	45	46	47	48	49	50

16. Janice sold 39 raffle tickets the first week. She sold 64 tickets the second week and 52 tickets the third week. How many tickets did she sell altogether?

17. Nick's car gets 27 miles to one gallon of gas. How many miles can he go on 100 gallons?

18. Star Machine Company made 3,567 parts in March. In April it made 5,409 parts. How many more parts did it make in April than in March?

19. The Lawsons are driving to San Diego. It is 3,025 miles from their home. They plan to make the trip in 5 days. How many miles must they drive each day?

20. Gloria makes $64 a day. How much does she make in 7 days?

UNIT TWO

7
UNDERSTANDING FRACTIONS AND MIXED NUMBERS

8
MULTIPLYING AND DIVIDING FRACTIONS

9
ADDING AND SUBTRACTING FRACTIONS

UNDERSTANDING FRACTIONS AND MIXED NUMBERS

7

Have you ever seen a listing of stock prices? What do the numbers look like? That's right, they're mixed numbers—whole numbers and fractions together. These people at the Pacific Stock Exchange trade stocks all day long. By day's end their heads must be filled with thousands of mixed numbers.

Chapter Learning Objectives

1. Explain fractions and their parts
2. Explain equal fractions
3. Change the terms of fractions
4. Find common denominators
5. Write improper fractions
6. Write mixed numbers
7. Change mixed numbers and improper fractions

WORDS TO KNOW

- **fraction** — a number that expresses part of a whole unit; $\frac{2}{3}$ is a fraction
- **denominator** — the bottom number in a fraction; the denominator tells how many equal parts there are in the whole unit
- **numerator** — the top number in a fraction; the numerator tells how many parts of the whole unit are being used
- **equivalent fractions** — fractions whose numbers are different but whose values are the same; $\frac{2}{8}$ and $\frac{1}{4}$ are equal fractions
- **like fractions** — fractions with the same denominator
- **unlike fractions** — fractions with different denominators
- **lowest terms** — a fraction is in its lowest terms when only 1 divides evenly into the numerator and denominator; it cannot be reduced
- **mixed number** — a number made up of a whole number and a fraction

7.1 What Are Fractions?

A **fraction** is part of a whole unit.

Numbers such as $\frac{1}{4}$, $\frac{1}{2}$, $\frac{2}{3}$, and $\frac{3}{4}$ are fractions.

Look at the square below. It's divided into 4 parts. Notice that 3 parts are colored.

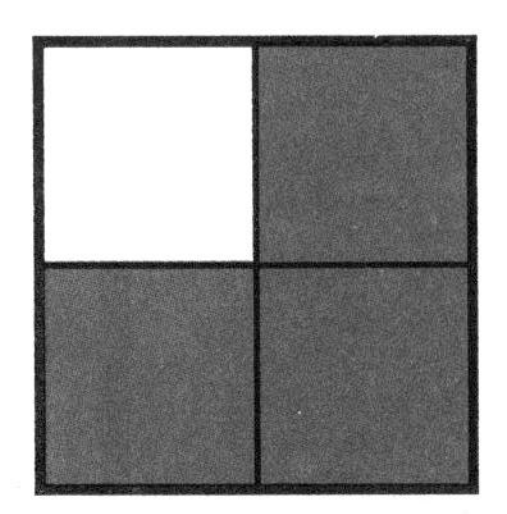

You can write a fraction that shows how many parts of the square are colored. Write the total number of parts as the bottom number, or denominator, of the fraction: $\frac{\ }{4}$. Write the number of colored parts as the top number, or numerator, of the fraction: $\frac{3}{4}$.

Every fraction has a numerator and a denominator.

The **denominator** is the bottom number. It tells how many parts there are in a whole unit.

The **numerator** is the top number. It tells how many parts of the unit are being used.

Example:
3 numerator
4 denominator

PRACTICE

On a separate sheet of paper, write a fraction to tell what part of each object is in color.

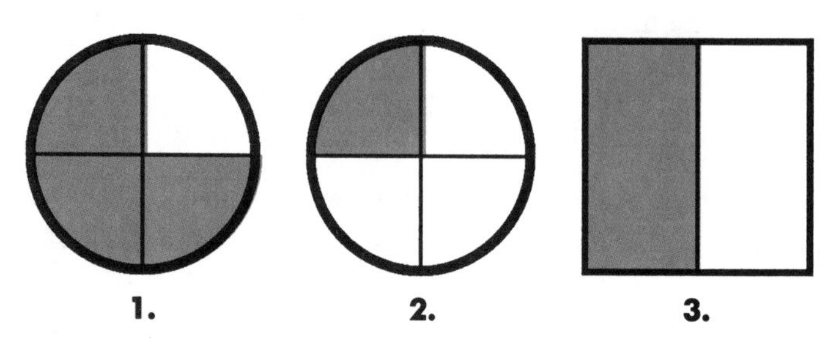

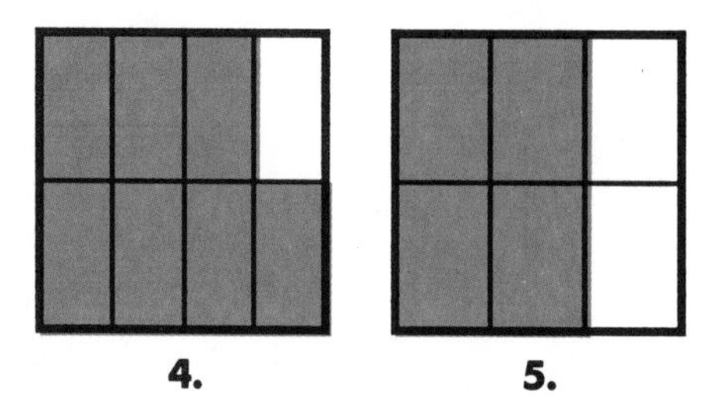

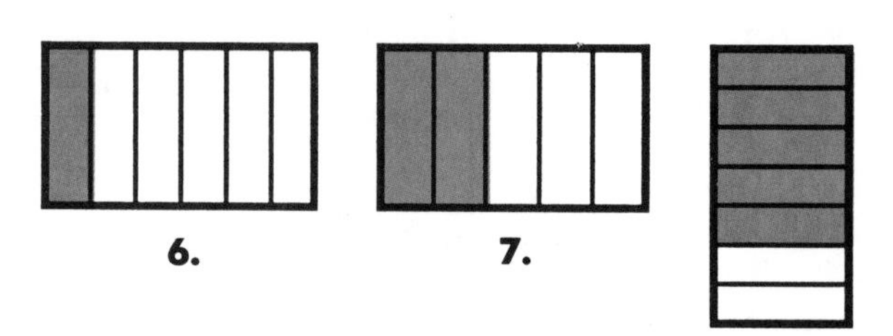

7.2 Recognizing Equivalent Fractions

Equivalent fractions have the same value even though their numbers are different.

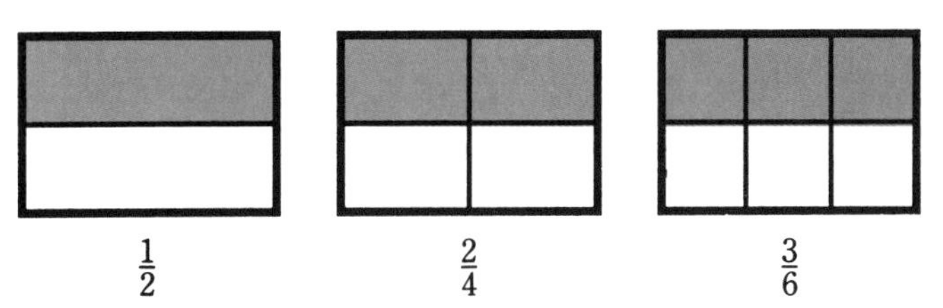

Each shape has been divided into a number of different parts. Look at the shapes. Think about these questions.

1. How many parts are in each shape?
2. How many parts of each shape are colored?

Each shape has a different number of parts. In each shape a different number of parts have been colored. But an equal area has been colored in each shape: the entire top half. The colored areas are all equal.

PRACTICE

On a separate piece of paper, write a fraction to tell how much of each drawing is in color. If equal amounts of both drawings are in color, write an equal sign. If the amounts in color are not equal, write an equal sign with a slash through it. The second sign means not equal.

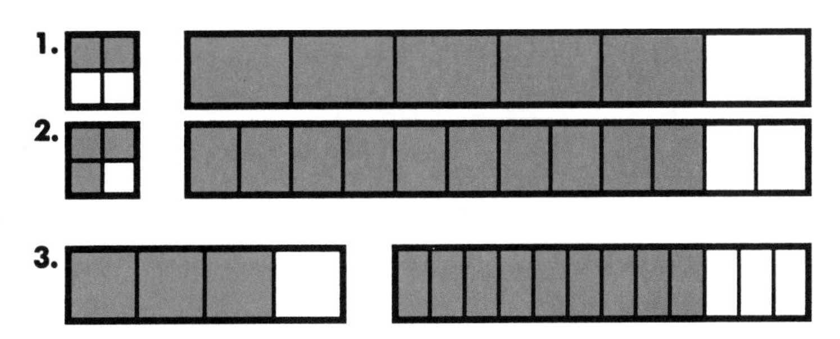

7.3 Changing Fractions to Higher Terms

Sometimes you need to express a fraction in higher terms. That is, you change the numbers of the fractions, but keep the value the same. For example:

Express $\frac{2}{3}$ as an equivalent fraction in higher terms with a denominator of 9.

$$\frac{2}{3} = \frac{?}{9}$$

Step 1: Divide the new denominator by the denominator of the given fraction.

$$9 \div 3 = 3$$

Step 2: Multiply the numerator and denominator of the given fraction by the quotient of Step 1.

Remember: the *quotient* is the answer in a division problem.

You find that $\frac{2}{3}$ can be expressed in higher terms as $\frac{6}{9}$.

$$\frac{2}{3} \times \frac{3}{3} = \frac{6}{9}$$

PRACTICE

Express the fractions below in higher terms. Use the new denominators to the right of each fraction. Do your work on a separate sheet of paper. The first one has been done for you.

1. $\frac{1}{2} = \frac{?}{8}$ $(8 \div 2 = 4)$ $\left(\frac{1}{2} \times \frac{4}{4} = \frac{4}{8}\right)$

2. $\frac{5}{9} = \frac{?}{18}$

3. $\frac{6}{10} = \frac{?}{30}$

4. $\frac{2}{9} = \frac{?}{27}$

5. $\frac{2}{7} = \frac{?}{35}$

6. $\frac{1}{3} = \frac{?}{21}$

7. $\frac{1}{12} = \frac{?}{60}$

8. $\frac{4}{9} = \frac{?}{36}$

9. $\frac{3}{7} = \frac{?}{42}$

10. $\frac{7}{9} = \frac{?}{45}$

11. $\frac{1}{5} = \frac{?}{25}$

12. $\frac{5}{7} = \frac{?}{49}$

13. $\frac{1}{4} = \frac{?}{20}$

14. $\frac{2}{8} = \frac{?}{24}$

15. $\frac{2}{3} = \frac{?}{15}$

16. $\frac{4}{6} = \frac{?}{36}$

17. $\frac{2}{5} = \frac{?}{50}$

18. $\frac{3}{15} = \frac{?}{45}$

19. $\frac{3}{8} = \frac{?}{32}$

20. $\frac{1}{6} = \frac{?}{54}$

7.4 Reducing Fractions to Lowest Terms

Sometimes you need to reduce fractions to their **lowest terms.** A fraction is in lowest terms when only 1 can be divided evenly into the numerator and the denominator. To do this, find the largest number that can divide evenly into the numerator and denominator. Divide the numerator and the denominator by this number.

Often, the largest number that can divide evenly into the numerator and denominator is the numerator. Try that number first.

Example A

$$\frac{3 \div 3}{6 \div 3} = \frac{1}{2}$$

Example B

$$\frac{4 \div 4}{12 \div 4} = \frac{1}{3}$$

In examples A and B, the numerator works. It divides evenly into both parts of the fraction.

But suppose dividing by the numerator doesn't work. For example, try reducing $\frac{6}{14}$ to its lowest terms:

$$\frac{6 \div 6}{14 \div 6} = \frac{1}{\text{doesn't divide evenly}}$$

When dividing by the numerator doesn't work, there is another way to reduce a fraction to its lowest terms. First, find the greatest common factor (GCF) of both the numerator and the denominator. Then divide the GCF into both these numbers. For example, here's how to reduce $\frac{6}{14}$ to its lowest terms:

If you need help finding the greatest common factor, turn back to page 105.

Step 1: List the common factors of both numbers.

$F_6 = \{1, 2, 3, 6\}$

$F_{14} = \{ 1, 2, 7, 14\}$

Step 2: Find the greatest common factor of both numbers. In this case, 2 is the GCF of 6 and 14.

Step 3: Divide the GCF into both parts of the fraction. The answer gives you the fraction in its lowest terms.

$$\frac{6 \div 2}{14 \div 2} = \frac{3}{7}$$

So, $\frac{6}{14}$ reduces to lowest terms of $\frac{3}{7}$.

Remember: A fraction is in its lowest terms when only 1 can be divided evenly into the numerator and denominator.

PRACTICE

Reduce the fractions below to their lowest terms. Do your work on a separate sheet of paper.

1. $\frac{4}{8}$

2. $\frac{5}{35}$

3. $\frac{9}{24}$

4. $\frac{8}{32}$

5. $\frac{3}{9}$

6. $\frac{12}{27}$

7. $\frac{6}{15}$

8. $\frac{6}{8}$

9. $\frac{5}{10}$

10. $\frac{7}{49}$

11. $\frac{13}{39}$

12. $\frac{16}{68}$

13. $\frac{8}{12}$

14. $\frac{8}{24}$

15. $\frac{49}{56}$

16. $\frac{6}{10}$

17. $\frac{10}{80}$

18. $\frac{6}{9}$

19. $\frac{24}{72}$

20. $\frac{7}{63}$

7.5 Finding Common Denominators

Like fractions have the *same* denominators.

Unlike fractions have *different* denominators.

You can change unlike fractions so that they have the same denominators.

Change $\frac{2}{3}$ and $\frac{5}{8}$ so that they have a common denominator.

Step 1: Find the least common multiple (LCM) of the denominators. The LCM of 3 and 8 is 24.

If you need help finding the least common multiple, turn back to page 104.

Step 2: Use 24 as the new denominator. $\frac{\ }{24}\ \frac{\ }{24}$

Step 3: Change $\frac{2}{3}$ to higher terms.
Divide the new denominator by the denominator of the given fraction.

$24 \div 3 = 8$

Multiply the numerator by the quotient.
$2 \times 8 = 16$

The new fraction is $\frac{16}{24}$. $\frac{16}{24} = \frac{2}{3}$

Step 4: Repeat Step 3 with the second fraction ($\frac{5}{8}$).

$24 \div 8 = 3$; $5 \times 3 = 15$. The new fraction is $\frac{15}{24}$. $\frac{15}{24} = \frac{5}{8}$

PRACTICE

A. Copy the fractions from each group that have common denominators. Do your work on a separate sheet of paper.

1. $\frac{2}{5}, \frac{2}{15}, \frac{3}{5}$

2. $\frac{3}{4}, \frac{6}{7}, \frac{1}{7}$

3. $\frac{22}{44}, \frac{21}{40}, \frac{2}{44}$

4. $\frac{7}{8}, \frac{2}{9}, \frac{4}{9}$

5. $\frac{3}{10}, \frac{2}{20}, \frac{6}{10}$

6. $\frac{3}{8}, \frac{3}{5}, \frac{1}{8}$

B. Change these unlike fractions to fractions with common denominators. Do your work on a separate sheet of paper.

1. $\frac{2}{3}, \frac{3}{4}$

2. $\frac{3}{9}, \frac{3}{4}$

3. $\frac{5}{12}, \frac{4}{7}$

4. $\frac{2}{13}, \frac{1}{3}$

5. $\frac{6}{8}, \frac{2}{7}$

6. $\frac{4}{5}, \frac{5}{6}$

7. $\frac{5}{8}, \frac{7}{12}$

8. $\frac{1}{2}, \frac{1}{9}$

9. $\frac{2}{7}, \frac{4}{9}$

10. $\frac{1}{4}, \frac{5}{6}$

11. $\frac{3}{8}, \frac{7}{10}$

12. $\frac{1}{2}, \frac{1}{11}$

13. $\frac{3}{15}, \frac{2}{7}$

14. $\frac{7}{12}, \frac{1}{5}$

15. $\frac{20}{24}, \frac{1}{6}$

16. $\frac{1}{7}, \frac{4}{5}$

17. $\frac{1}{3}, \frac{5}{9}$

18. $\frac{3}{4}, \frac{1}{14}$

19. $\frac{8}{9}, \frac{2}{15}$

20. $\frac{2}{18}, \frac{1}{10}$

7.6 Comparing Fractions

Comparing fractions with the same denominator is simple. Just compare the numerators. The fraction with the larger numerator is the larger fraction. For example: $\frac{6}{8}$ is larger than $\frac{2}{8}$, because 6 is larger than 2.

You can also compare fractions with *different* denominators.

Compare $\frac{2}{5}$ and $\frac{3}{4}$. Follow the steps below.

Step 1: Find a common denominator.

The common denominator of these fractions is 20.

Remember: the common denominator is the LCM of both denominators of the fractions.

Step 2: Express both fractions in higher terms. Remember to use the *new* common denominator.

$\frac{2}{5} = \frac{8}{20}$ $\frac{3}{4} = \frac{15}{20}$

Step 3: Compare the numerators.

15 is larger than 8. Therefore, $\frac{3}{4}$ is a larger fraction than $\frac{2}{5}$.

PRACTICE

Compare each pair of fractions below. Write the larger fraction in each pair on a separate sheet of paper.

1. $\frac{2}{3}, \frac{3}{4}$

2. $\frac{3}{8}, \frac{2}{7}$

3. $\frac{5}{11}, \frac{7}{12}$

4. $\frac{5}{6}, \frac{3}{4}$

5. $\frac{9}{11}, \frac{2}{5}$

6. $\frac{1}{5}, \frac{1}{4}$

7. $\frac{2}{3}, \frac{6}{7}$

8. $\frac{1}{5}, \frac{2}{7}$

9. $\frac{5}{8}, \frac{4}{5}$

10. $\frac{2}{9}, \frac{1}{8}$

11. $\frac{3}{11}, \frac{1}{3}$

12. $\frac{1}{2}, \frac{4}{7}$

13. $\frac{3}{10}, \frac{1}{3}$

14. $\frac{1}{4}, \frac{2}{5}$

15. $\frac{3}{4}, \frac{7}{9}$

16. $\frac{7}{8}, \frac{4}{5}$

17. $\frac{1}{3}, \frac{2}{5}$

18. $\frac{5}{12}, \frac{3}{5}$

19. $\frac{1}{6}, \frac{2}{9}$

20. $\frac{5}{6}, \frac{9}{10}$

7.7 Recognizing Mixed Numbers and Improper Fractions

A **mixed number** is a number that is made up of a whole number and a fraction.

The following are mixed numbers: $1\frac{1}{2}$, $2\frac{4}{5}$, $13\frac{2}{3}$

A *proper fraction* is a fraction whose numerator is smaller than its denominator.

These are proper fractions: $\frac{1}{2}$, $\frac{2}{3}$, $\frac{7}{11}$, $\frac{12}{15}$

An *improper fraction* is a fraction whose numerator is larger than its denominator.

These are improper fractions: $\frac{2}{1}$, $\frac{3}{2}$, $\frac{11}{7}$, $\frac{15}{12}$

PRACTICE

Number a separate sheet of paper from 1 to 20. Decide whether each of the following numbers are proper fractions, improper fractions, or mixed numbers. Write *PF* for proper fractions, *IF* for improper fractions, or *MN* for mixed numbers.

1. $\frac{3}{4}$
2. $\frac{8}{5}$
3. $3\frac{2}{5}$
4. $6\frac{7}{9}$
5. $\frac{9}{4}$
6. $\frac{11}{12}$
7. $\frac{5}{4}$
8. $11\frac{2}{3}$
9. $\frac{22}{7}$
10. $\frac{6}{9}$
11. $15\frac{7}{8}$
12. $\frac{7}{5}$
13. $\frac{4}{3}$
14. $4\frac{1}{10}$
15. $\frac{17}{15}$
16. $5\frac{4}{5}$
17. $2\frac{1}{20}$
18. $\frac{8}{9}$
19. $7\frac{4}{9}$
20. $\frac{19}{15}$
21. $\frac{3}{8}$
22. $3\frac{3}{8}$
23. $\frac{11}{5}$
24. $\frac{1}{2}$
25. $\frac{9}{8}$
26. $1\frac{1}{4}$
27. $\frac{5}{9}$
28. $7\frac{2}{3}$
29. $\frac{18}{16}$
30. $\frac{5}{6}$

7.8 Changing Improper Fractions to Mixed Numbers

Improper fractions can be changed to mixed numbers.

Example: Change $\frac{34}{8}$ to a mixed number. Follow the steps below.

Step 1: Divide the numerator by the denominator.
$34 \div 8 = 4 \text{ R } 2$

Step 2: Write the quotient as a mixed number. Express the remainder as a fraction. The mixed number is $4\frac{2}{8}$.

Step 3: If the fraction can be reduced, reduce it.
$\frac{34}{8} = 4\frac{2}{8} = 4\frac{1}{4}$

Some improper fractions can be changed to whole numbers.

Change $\frac{32}{8}$ to a whole number. Follow the steps below.

Step 1: Divide the numerator by the denominator.
$32 \div 8 = 4$

Step 2: Write the quotient as a whole number. 4

PRACTICE

A. Change the improper fractions below to mixed or whole numbers. Do your work on a separate sheet of paper.

1. $\frac{17}{9}$	**5.** $\frac{18}{3}$	**9.** $\frac{42}{6}$	**13.** $\frac{14}{11}$	**17.** $\frac{25}{3}$
2. $\frac{8}{5}$	**6.** $\frac{22}{7}$	**10.** $\frac{20}{4}$	**14.** $\frac{70}{12}$	**18.** $\frac{82}{9}$
3. $\frac{10}{3}$	**7.** $\frac{54}{8}$	**11.** $\frac{80}{9}$	**15.** $\frac{35}{5}$	**19.** $\frac{63}{7}$
4. $\frac{16}{8}$	**8.** $\frac{29}{3}$	**12.** $\frac{65}{3}$	**16.** $\frac{9}{4}$	**20.** $\frac{56}{6}$

B. Write five improper fractions of your own. Then change them to mixed or whole numbers. Do your work on a separate sheet of paper.

7.9 Changing Mixed Numbers to Improper Fractions

Mixed numbers can be changed to improper fractions.

Example: Change $9\frac{2}{3}$ to an improper fraction. Follow the steps below.

Step 1: Multiply the whole number by the denominator of the fraction.

$9 \times 3 = 27$

Step 2: Add the numerator to the product. $27 + 2 = 29$

Step 3: Use this sum as the numerator of the improper fraction. The denominator stays the same.
$9\frac{2}{3} = \frac{29}{3}$

PRACTICE

A. Change the mixed numbers below to improper fractions. Do your work on a separate sheet of paper.

1. $4\frac{3}{4}$

2. $5\frac{2}{5}$

3. $7\frac{1}{8}$

4. $3\frac{6}{7}$

5. $9\frac{4}{5}$

6. $6\frac{7}{12}$

7. $8\frac{5}{9}$

8. $7\frac{3}{4}$

9. $2\frac{17}{18}$

10. $8\frac{2}{3}$

11. $12\frac{4}{6}$

12. $5\frac{8}{9}$

B. Write 5 mixed numbers of your own. Change them to improper fractions. Do your work on a separate sheet of paper.

MATHEMATICS IN YOUR LIFE:
Using a Ruler

Jim just moved into a new apartment. He needs to find out how his furniture will fit in the new rooms. The strips below show the different lengths of some of Jim's furniture.

Measure each strip with a ruler. Suppose that one inch equals one foot, two inches equals two feet, and so on. Next, look at the chart below the strips. Then answer the questions on the next page. Write on a separate sheet of paper.

Bookcase

Couch

Dresser

Stove

Refrigerator

Room	Measurement
Bedroom	Longest wall - 10 feet
Living Room	Longest wall - 15 feet
Kitchen	Longest wall - 8 feet

1. Can Jim fit a bookcase and dresser side-by-side in his bedroom?
2. Can Jim put two more bookcases next to his couch along one wall in the living room?
3. Will the refrigerator and stove fit next to each other in Jim's kitchen? Will there be any room left over? How much?

7.10 Using Your Calculator: Changing Mixed Numbers to Fractions

You can use your calculator to change mixed numbers to fractions.

Change $32\frac{3}{5}$ to a fraction.

Step 1: Press [3] [2]

Step 2: Press [×]

Step 3: Press [5]

Step 4: Press [=]

Step 5: Press [+]

Step 6: Press [3]

Step 7: Press [=]

Step 8: Your answer should be: $32\frac{3}{5} = \frac{163}{5}$

PRACTICE

Use your calculator to change these mixed numbers to improper fractions.

1. $39\frac{4}{5} =$

2. $74\frac{7}{8} =$

3. $67\frac{8}{9} =$

4. $25\frac{1}{4} =$

5. $82\frac{2}{3} =$

6. $91\frac{6}{7} =$

7. $59\frac{1}{2}$

8. $43\frac{1}{6}$

CHAPTER SUMMARY

■ Fraction	A fraction is a number which expresses part of a whole. Every fraction has a numerator and a denominator.
■ Denominator	The denominator is the bottom number of a fraction. The denominator tells how many equal parts there are in a whole unit.
■ Numerator	The numerator is the top number in a fraction. The numerator tells how many parts of a whole unit are being used.
■ Equivalent Fractions	Equivalent fractions have different numbers, but the same value.
■ Like and Unlike Fractions	Like fractions have the same denominator. Unlike fractions have different denominators.
■ Lowest Terms	A fraction has been reduced to its lowest terms when only 1 can be divided evenly into both the numerator and denominator.
■ Mixed Number	A mixed number is made up of a whole number and a fraction.

CHAPTER QUIZ

A. Compare each pair of fractions below. Write the larger fraction in each pair on a separate sheet of paper.

1. $\frac{2}{3}, \frac{7}{8}$ **3.** $\frac{1}{2}, \frac{3}{5}$ **5.** $\frac{3}{10}, \frac{1}{3}$ **7.** $\frac{4}{9}, \frac{5}{7}$ **9.** $\frac{5}{6}, \frac{8}{9}$

2. $\frac{4}{5}, \frac{3}{7}$ **4.** $\frac{3}{8}, \frac{5}{6}$ **6.** $\frac{3}{4}, \frac{1}{6}$ **8.** $\frac{5}{8}, \frac{7}{9}$ **10.** $\frac{7}{10}, \frac{4}{5}$

B. Change each improper fraction below to a mixed number.

1. $\frac{3}{2}$ **3.** $\frac{4}{3}$ **5.** $\frac{15}{13}$ **7.** $\frac{7}{4}$ **9.** $\frac{11}{8}$

2. $\frac{8}{5}$ **4.** $\frac{7}{3}$ **6.** $\frac{12}{5}$ **8.** $\frac{9}{8}$ **10.** $\frac{19}{6}$

CHAPTER REVIEW

CHAPTER QUIZ

C. Change each mixed number below to an improper fraction.

1. $15\frac{1}{2}$ **5.** $4\frac{3}{5}$ **9.** $3\frac{5}{11}$ **13.** $5\frac{3}{4}$

2. $6\frac{1}{20}$ **6** $11\frac{5}{7}$ **10.** $7\frac{3}{10}$ **14.** $8\frac{3}{8}$

3. $17\frac{2}{3}$ **7.** $12\frac{2}{3}$ **11.** $9\frac{2}{9}$ **15.** $10\frac{4}{9}$

4. $20\frac{1}{4}$ **8.** $7\frac{3}{9}$ **12.** $13\frac{1}{3}$ **16.** $13\frac{4}{7}$

D. Reduce these fractions to their lowest terms.

1. $\frac{20}{24}$ **2.** $\frac{8}{16}$ **3.** $\frac{10}{50}$ **4.** $\frac{4}{18}$ **5.** $\frac{6}{40}$

REVIEWING VOCABULARY

Use the correct word or words from the box to complete the sentences. Write on a separate sheet of paper.

fraction	denominator	mixed number
numerator		equivalent fractions

1. The ________ is the bottom number in a fraction.

2. The ________ is the top number in a fraction.

3. ________ have the same value.

4. A ________ is a number which expresses part of a whole unit.

5. A ________ is made up of a whole number and a fraction.

MULTIPLYING AND DIVIDING FRACTIONS

8

In order to cook from a recipe, you need to know how to measure. To double a recipe or cut it in half, you also need to know how to multiply and divide fractions.

Chapter Learning Objectives

1. Write proper and improper fractions
2. Write mixed numbers
3. Multiply fractions
4. Multiply mixed numbers
5. Divide fractions
6. Divide mixed numbers
7. Solve problems with fractions and mixed numbers

WORDS TO KNOW

- **cross cancel** to divide a numerator and a denominator by the same number
- **invert** to reverse the positions of the numerator and denominator of a fraction

8.1 Multiplying Fractions

Multiplying fractions is a simple process. If you can multiply whole numbers, then you can multiply fractions.

For example: multiply $\frac{3}{4} \times \frac{2}{5}$

Follow the steps below.

Step 1: Multiply the numerators. $\frac{3}{4} \times \frac{2}{5} = \frac{6}{\ }$

Step 2: Multiply the denominators. $\frac{3}{4} \times \frac{2}{5} = \frac{6}{20}$

Step 3: Reduce the product to lowest terms. $\frac{6}{20} = \frac{3}{10}$

PRACTICE

Solve the problems below on a separate sheet of paper. Reduce the products to the lowest terms whenever possible.

1. $\frac{3}{8} \times \frac{3}{4} =$

2. $\frac{5}{6} \times \frac{1}{2} =$

3. $\frac{2}{3} \times \frac{4}{9} =$

4. $\frac{1}{8} \times \frac{5}{8} =$

5. $\frac{2}{9} \times \frac{6}{11} =$

6. $\frac{1}{5} \times \frac{4}{7} =$

7. $\frac{5}{8} \times \frac{11}{12} =$

8. $\frac{7}{9} \times \frac{13}{15} =$

9. $\frac{4}{7} \times \frac{3}{4} =$

10. $\frac{3}{5} \times \frac{5}{8} =$

11. $\frac{1}{3} \times \frac{5}{16} =$

12. $\frac{7}{9} \times \frac{2}{3} =$

13. $\frac{3}{4} \times \frac{9}{10} =$

14. $\frac{1}{12} \times \frac{2}{3} =$

15. $\frac{2}{15} \times \frac{1}{2} =$

16. $\frac{7}{9} \times \frac{8}{10} =$

17. $\frac{4}{5} \times \frac{1}{2} =$

18. $\frac{6}{8} \times \frac{7}{9} =$

19. $\frac{3}{11} \times \frac{2}{3} =$

20. $\frac{5}{12} \times \frac{5}{6} =$

21. $\frac{3}{5} \times \frac{2}{3} =$

22. $\frac{4}{7} \times \frac{7}{8} =$

23. $\frac{3}{8} \times \frac{4}{9} =$

24. $\frac{2}{3} \times \frac{2}{5} =$

25. $\frac{5}{6} \times \frac{3}{10} =$

26. $\frac{1}{2} \times \frac{4}{7} =$

27. $\frac{4}{5} \times \frac{5}{8} =$

28. $\frac{2}{5} \times \frac{5}{8} =$

8.2 Cross Cancellation

Cross cancellation is a method that makes multiplying fractions easier. **Cancelling** is simply dividing the numerator of one fraction and the denominator of another fraction by the same number. Cancel fractions whenever you can before multiplying.

Multiply $\frac{9}{14} \times \frac{4}{27}$. Use cross cancellation.

Step 1: Divide 9 into the numerator 9. Cross out 9 and write 1.

Divide 9 into the denominator 27. Cross out 27 and write 3.

$$\frac{\overset{1}{\cancel{9}}}{14} \times \frac{4}{\underset{3}{\cancel{27}}} =$$

Step 2: Divide 2 into the numerator 4. Cross out 4 and write 2.

Divide 2 into the denominator 14. Cross out 14 and write 7.

$$\frac{\overset{1}{\cancel{9}}}{\underset{7}{\cancel{14}}} \times \frac{\overset{2}{\cancel{4}}}{\underset{3}{\cancel{27}}} =$$

Step 3: Multiply the new numerators. $1 \times 2 = 2$

Step 4: Multiply the new denominators. $7 \times 3 = 21$

$$\frac{\overset{1}{\cancel{9}}}{\underset{7}{\cancel{14}}} \times \frac{\overset{2}{\cancel{4}}}{\underset{3}{\cancel{27}}} = \frac{2}{21}$$

Answer: $\frac{2}{21}$

PRACTICE

Solve the problems below on a separate sheet of paper. Use cross cancellation to make multiplying easier. Reduce the products to lowest terms whenever possible.

1. $\frac{2}{3} \times \frac{6}{7} =$

2. $\frac{2}{5} \times \frac{10}{16} =$

3. $\frac{3}{4} \times \frac{2}{9} =$

4. $\frac{4}{7} \times \frac{14}{20} =$

5. $\frac{5}{6} \times \frac{12}{25} =$

6. $\frac{4}{7} \times \frac{1}{4} =$

7. $\frac{2}{6} \times \frac{9}{10} =$

8. $\frac{4}{5} \times \frac{5}{18} =$

9. $\frac{3}{10} \times \frac{8}{21} =$

10. $\frac{14}{16} \times \frac{4}{7} =$

11. $\frac{5}{9} \times \frac{4}{30} =$

12. $\frac{9}{32} \times \frac{8}{9} =$

13. $\frac{1}{8} \times \frac{6}{7} =$

14. $\frac{4}{5} \times \frac{1}{3} =$

15. $\frac{7}{8} \times \frac{16}{49} =$

16. $\frac{9}{12} \times \frac{1}{27} =$

17. $\frac{7}{9} \times \frac{18}{19} =$

18. $\frac{2}{5} \times \frac{1}{2} =$

19. $\frac{4}{9} \times \frac{3}{28} =$

20. $\frac{1}{4} \times \frac{44}{51} =$

8.3 Multiplying Mixed Numbers

Before multiplying mixed numbers, you have to change them to improper fractions.

Remember: an improper fraction is a fraction whose numerator is larger than its denominator.

For example: Multiply $5\frac{1}{3} \times 2\frac{1}{2}$

Follow the steps below.

If you need help changing mixed numbers to improper fractions, turn back to page 123.

Step 1: Change each mixed number to an improper fraction.

$\frac{16}{3} \times \frac{5}{2} =$

Step 2: Multiply. Use cross cancellation whenever you can.

$\frac{\overset{8}{\cancel{16}}}{3} \times \frac{5}{\underset{1}{\cancel{2}}} = \frac{40}{3}$

Step 3: Change the improper fraction to a mixed number.

$\frac{40}{3} = 13\frac{1}{3}$

PRACTICE

Solve the problems below on a separate sheet of paper. Use cross cancellation to make multiplying easier. Reduce the products to lowest terms whenever possible.

1. $3\frac{3}{4} \times 2\frac{1}{8} =$

2. $4\frac{1}{2} \times 7\frac{1}{3} =$

3. $5\frac{6}{7} \times 2\frac{4}{9} =$

4. $8\frac{2}{3} \times 2\frac{4}{12} =$

5. $9\frac{7}{10} \times 3\frac{1}{3} =$

6. $6\frac{2}{3} \times \frac{2}{8} =$

7. $\frac{7}{9} \times 1\frac{1}{2} =$

8. $8\frac{1}{3} \times \frac{3}{4} =$

9. $4\frac{5}{6} \times 5\frac{2}{5} =$

10. $7\frac{4}{7} \times \frac{7}{8} =$

11. $\frac{4}{5} \times 2\frac{9}{10} =$

12. $5\frac{7}{8} \times 6\frac{1}{9} =$

13. $3\frac{5}{8} \times \frac{2}{5} =$

14. $3\frac{1}{2} \times 2\frac{3}{4} =$

15. $\frac{4}{5} \times 2\frac{1}{2} =$

16. $4\frac{7}{8} \times 3\frac{1}{4} =$

17. $6\frac{1}{3} \times 2\frac{2}{5} =$

18. $\frac{1}{4} \times 8\frac{2}{4} =$

19. $1\frac{1}{4} \times \frac{4}{5} =$

20. $5\frac{3}{5} \times 2\frac{1}{2} =$

8.4 Multiplying Fractions and Whole Numbers

To multiply a fraction and a whole number, first change the whole number to an improper fraction.

You can change any whole number to an improper fraction. Make the whole number the numerator. Make the denominator 1. $2 = \frac{2}{1}$ $16 = \frac{16}{1}$ $39 = \frac{39}{1}$

Multiply $8 \times \frac{2}{3}$

Step 1: Change the whole number to an improper fraction. $8 = \frac{8}{1}$

Step 2: Multiply. $\frac{8}{1} \times \frac{2}{3} = \frac{16}{3}$

Step 3: Change the product to a mixed number. $\frac{16}{3} = 5\frac{1}{3}$

PRACTICE

A. Solve the problems below on a separate sheet of paper. Change any whole number to an improper fraction. Reduce the products to lowest terms.

1. $2 \times \frac{3}{4} =$

2. $\frac{7}{8} \times 5 =$

3. $\frac{3}{4} \times 8 =$

4. $\frac{4}{5} \times 40 =$

5. $\frac{3}{10} \times 17 =$

6. $15 \times \frac{2}{3} =$

7. $4 \times \frac{8}{9} =$

8. $7 \times \frac{3}{5} =$

9. $\frac{2}{3} \times 9 =$

10. $\frac{5}{7} \times 12 =$

11. $81 \times \frac{5}{9} =$

12. $52 \times \frac{3}{8} =$

13. $74 \times \frac{5}{12} =$

14. $\frac{2}{14} \times 38 =$

15. $\frac{5}{6} \times 11 =$

16. $32 \times \frac{7}{8} =$

17. $\frac{3}{11} \times 44 =$

18. $64 \times \frac{3}{8} =$

19. $\frac{2}{7} \times 84 =$

20. $\frac{5}{9} \times 2 =$

B. Solve the problems below on a separate sheet of paper. Change any mixed number or whole number to an improper fraction. Reduce the products to lowest terms.

1. $3\frac{2}{3} \times \frac{4}{9} =$

2. $8 \times \frac{5}{7} =$

3. $\frac{2}{3} \times \frac{7}{11} =$

4. $2\frac{6}{8} \times 4\frac{1}{4} =$

5. $7\frac{1}{3} \times 21 =$

6. $\frac{4}{5} \times \frac{9}{11} =$

7. $3\frac{5}{8} \times 24 =$

8. $8 \times 7\frac{1}{6} =$

9. $2\frac{3}{4} \times 6 =$

10. $5\frac{7}{8} \times \frac{3}{5} =$

11. $7 \times 2\frac{4}{5} =$

12. $33 \times \frac{7}{11} =$

13. $45 \times 3\frac{2}{15} =$

14. $86 \times 4\frac{5}{8} =$

15. $12\frac{8}{10} \times \frac{1}{2} =$

16. $5\frac{3}{4} \times 5 =$

17. $42 \times 3\frac{1}{7} =$

18. $9 \times \frac{1}{9} =$

19. $27 \times 5\frac{2}{3} =$

20. $32\frac{3}{4} \times 1 =$

8.5 Dividing a Fraction by a Fraction

Dividing a fraction by another fraction is not difficult. But before you can proceed, you need to make certain changes in the problem. You will have to **invert** the fraction. For example: $\frac{3}{4} \div \frac{4}{6}$

Follow the steps below.

Step 1: *Invert* the second fraction. Make the numerator the denominator. Make the denominator the numerator.

$$\frac{3}{4} \div \frac{6}{4} =$$

Step 2: Change the division sign to a multiplication sign.

$$\frac{3}{4} \times \frac{6}{4} =$$

Step 3: Cancel if possible.

Step 4: *Multiply* the numerators.

Step 5: *Multiply* the denominators.

$$\frac{3}{{}_2\not{4}} \times \frac{\not{6}^3}{4} = \frac{9}{8}$$

Step 6: If the product is an improper fraction, change it to a mixed number.

$$\frac{9}{8} = 1\frac{1}{8}$$

PRACTICE

Solve the problems below on a separate sheet of paper.

1. $\frac{2}{3} \div \frac{2}{8} =$

2. $\frac{3}{4} \div \frac{1}{8} =$

3. $\frac{5}{9} \div \frac{1}{2} =$

4. $\frac{7}{9} \div \frac{1}{3} =$

5. $\frac{4}{5} \div \frac{1}{15} =$

6. $\frac{1}{2} \div \frac{1}{4} =$

7. $\frac{7}{8} \div \frac{1}{3} =$

8. $\frac{5}{6} \div \frac{1}{5} =$

9. $\frac{3}{5} \div \frac{7}{8} =$

10. $\frac{6}{11} \div \frac{1}{2} =$

11. $\frac{3}{5} \div \frac{5}{8} =$

12. $\frac{11}{16} \div \frac{1}{4} =$

13. $\frac{2}{3} \div \frac{1}{3} =$

14. $\frac{7}{12} \div \frac{1}{5} =$

15. $\frac{12}{13} \div \frac{4}{5} =$

16. $\frac{9}{24} \div \frac{2}{3} =$

17. $\frac{4}{5} \div \frac{2}{5} =$

18. $\frac{9}{10} \div \frac{6}{7} =$

19. $\frac{4}{10} \div \frac{1}{5} =$

20. $\frac{12}{20} \div \frac{1}{2} =$

8.6 Dividing a Fraction by a Whole Number

The method for dividing a fraction by a whole number is similar to dividing a fraction by a fraction. For example: $\frac{7}{8} \div 5$

Follow the steps below.

Step 1: Change the whole number to an improper fraction. $\frac{7}{8} \div \frac{5}{1} =$

Step 2: Invert the second fraction. $\frac{7}{8} \times \frac{1}{5} =$

Step 3: Change the sign to multiplication.

Step 4: Multiply the numerators and denominators. $\frac{7}{8} \times \frac{1}{5} = \frac{7}{40}$

> The above steps also apply when you want to divide a whole number by a fraction. And don't forget to cancel fractions whenever possible.

PRACTICE

Solve the problems below on a separate sheet of paper.

1. $\frac{2}{3} \div 3 =$

2. $\frac{1}{2} \div 4 =$

3. $\frac{3}{4} \div 5 =$

4. $3 \div \frac{7}{8} =$

5. $25 \div \frac{1}{5} =$

6. $\frac{1}{2} \div 7 =$

7. $\frac{3}{4} \div 18 =$

8. $\frac{4}{5} \div 20 =$

9. $24 \div \frac{5}{12} =$

10. $32 \div \frac{6}{7} =$

11. $\frac{7}{9} \div 6 =$

12. $52 \div \frac{2}{3} =$

13. $49 \div \frac{7}{10} =$

14. $10 \div \frac{1}{8} =$

15. $\frac{1}{3} \div 10 =$

16. $6 \div \frac{3}{4} =$

17. $\frac{3}{4} \div 6 =$

18. $\frac{5}{7} \div 63 =$

19. $\frac{8}{9} \div 24 =$

20. $\frac{10}{11} \div 20 =$

21. $18 \div \frac{2}{3} =$

22. $\frac{5}{6} \div 5 =$

23. $20 \div \frac{4}{5} =$

24. $16 \div \frac{1}{2} =$

25. $\frac{3}{8} \div 12 =$

26. $30 \div \frac{5}{6} =$

27. $\frac{5}{6} \div 25 =$

28. $75 \div \frac{3}{25} =$

8.7 Dividing a Fraction by a Mixed Number

Here is the method for dividing a fraction by a mixed number.

For example: $\frac{7}{9} \div 5\frac{2}{3}$

Follow the steps below.

Step 1: Change the mixed number to an improper fraction.

$$\frac{7}{9} \div \frac{17}{3} =$$

Step 2: Invert the second fraction.

$$\frac{7}{9} \div \frac{3}{17} =$$

Step 3: Change the sign to multiplication.

Step 4: Multiply the numerators and denominators.

$$\frac{7}{9} \times \frac{3}{17} = \frac{21}{153}$$

Step 5: Reduce to lowest terms if possible.

$$\frac{21}{153} = \frac{7}{51}$$

PRACTICE

Solve the problems below on a separate sheet of paper.

1. $\frac{2}{3} \div 1\frac{2}{3} =$
2. $4\frac{2}{5} \div \frac{2}{5} =$
3. $6\frac{1}{8} \div 3\frac{1}{2} =$
4. $3\frac{1}{8} \div \frac{7}{8} =$
5. $4\frac{5}{8} \div 2\frac{1}{3} =$
6. $7\frac{1}{7} \div 2\frac{2}{3} =$
7. $\frac{17}{18} \div 2\frac{1}{2} =$
8. $\frac{4}{5} \div 3\frac{3}{4} =$
9. $7\frac{9}{10} \div \frac{1}{5} =$
10. $2\frac{1}{9} \div 2\frac{1}{10} =$
11. $\frac{7}{8} \div 4\frac{2}{5} =$
12. $6\frac{1}{2} \div 5\frac{7}{9} =$
13. $11\frac{3}{8} \div 4\frac{1}{2} =$
14. $15\frac{2}{6} \div 6\frac{7}{8} =$
15. $20\frac{1}{2} \div 9\frac{1}{3} =$
16. $5\frac{6}{16} \div 3\frac{1}{8} =$
17. $\frac{6}{19} \div 2\frac{8}{19} =$
18. $\frac{4}{5} \div 3\frac{2}{7} =$
19. $\frac{8}{14} \div 6\frac{1}{7} =$
20. $\frac{7}{12} \div 4\frac{1}{6} =$
21. $3\frac{7}{8} \div 2\frac{1}{3} =$
22. $\frac{3}{5} \div 2\frac{1}{3} =$
23. $4\frac{1}{5} \div \frac{2}{5} =$
24. $3\frac{3}{8} \div 3\frac{3}{8} =$
25. $4\frac{3}{4} \div \frac{1}{2} =$
26. $2\frac{5}{6} \div \frac{1}{6} =$
27. $\frac{3}{8} \div 2\frac{1}{4} =$
28. $3\frac{7}{8} \div 1\frac{1}{2} =$

8.8 Solving Problems with Fractions

Steps in Solving Word Problems

Read this problem. Follow the steps below to solve it.

Rita had $\frac{3}{4}$ of a yard of cloth. She used $\frac{1}{2}$ of it. How much cloth did she have left?

Step 1: Read the problem.

Step 2: Learn what you must find out.

How much cloth did Rita have left?

Step 3: Notice the clue words. Clue words help you solve the problem. In this problem, there is one clue word, *of*. *Of* suggests that you may have to multiply.

Clue word: of

Step 4: Find the numbers.

$\frac{3}{4}$ $\frac{1}{2}$

Step 5: Write the problem. Solve it.

$\frac{3}{4} \times \frac{1}{2} = \frac{3}{8}$

Step 6: Check the answer. Ask yourself if it makes sense.

Clue Words

For multiplication: of; part
For division: how many; in
Remember: Not all problems have clue words.

PROBLEMS TO SOLVE

Carefully read the problems below. Solve them on a separate sheet of paper. If your answer is an improper fraction, change it to a mixed number. Make sure your answer is in its lowest terms.

Multiplication Problems:

1. A total of 120 people stood in line for the movie. There were seats for $\frac{3}{4}$ of them. How many of the people could be seated?
2. Julie's recipe for cookies calls for $2\frac{1}{4}$ cups of flour. Julie wants to make $3\frac{1}{2}$ times as many cookies as the recipe calls for. How much flour will she need?
3. Jason painted a room in $2\frac{1}{5}$ hours. How long would it take him to paint four rooms?
4. It took Bob $6\frac{3}{4}$ hours to repair his car. Dave said that he could have done it in half the time. How long would it have taken Dave?
5. Gloria made 96 cookies for a bake sale. She sold $\frac{2}{3}$ of them. How many of them did she sell?

Division Problems:

6. The bookshelf is 42 inches long. Each book is $1\frac{1}{4}$ inches wide. How many books will fit on the shelf?
7. A strip of wallpaper is $2\frac{1}{2}$ feet wide. How many strips are needed to cover a wall 8 feet wide?
8. Jessica bought 9 yards of fabric. If one apron uses $2\frac{1}{3}$ yards of fabric, how many aprons can she make?
9. Marigolds need to be planted $2\frac{3}{4}$ inches apart. How many plants should you buy for one row 18 inches long?
10. If one shoe box measures $6\frac{1}{2}$ inches across, how many shoe boxes will fit on a shelf 48 inches long?

MATHEMATICS IN YOUR LIFE: Overtime Pay

The employees of Apex Company work 40 hours a week. They are paid overtime for every hour over 40 that they work. The overtime rate is $1\frac{1}{2}$ times the regular hourly rate.

Janice earns eight dollars an hour.

Here is how much she makes for four hours of overtime.

$$\$8. \times 1\frac{1}{2}$$

$$\frac{8}{1} \times \frac{3}{2} = \frac{24}{2} = 12$$

$$\begin{array}{r} \$\ 12.00 \\ \times \quad 4 \\ \hline \$\ 48.00 \end{array}$$

PRACTICE

Copy this chart on a separate sheet of paper. Figure out how much money each worker made in overtime pay. Add these amounts to the chart.

Employee	Regular Rate	Overtime Hours	Overtime Pay
Wu, Laura	$9.00	3	
Reynolds, John	$7.00	2	
West, Linda	$8.00	$1\frac{1}{2}$	
Alvarez, Dan	$6.00	6	
Ellison, Mark	$6.00	$3\frac{1}{2}$	
Takahashi, Michael	$8.00	$4\frac{1}{2}$	
Phillips, Ann	$7.00	8	
Ferraro, Stanley	$9.00	$5\frac{1}{4}$	

8.10 Using Your Calculator: Multiplying Fractions

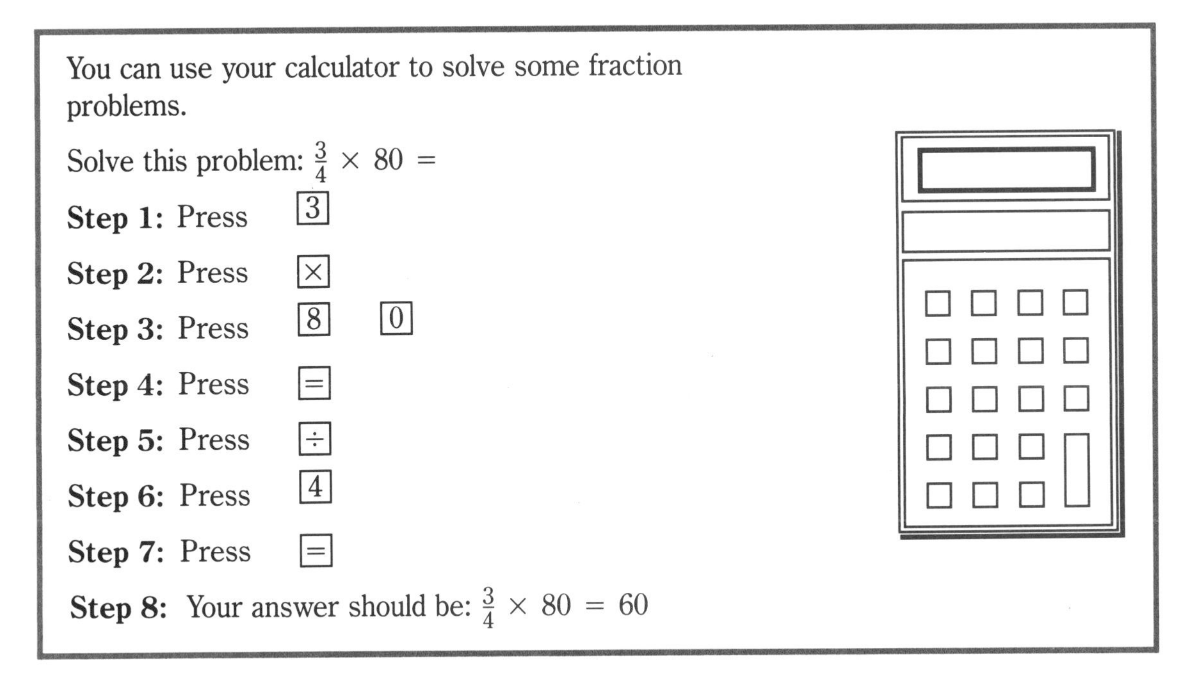

You can use your calculator to solve some fraction problems.

Solve this problem: $\frac{3}{4} \times 80 =$

Step 1: Press [3]

Step 2: Press [×]

Step 3: Press [8] [0]

Step 4: Press [=]

Step 5: Press [÷]

Step 6: Press [4]

Step 7: Press [=]

Step 8: Your answer should be: $\frac{3}{4} \times 80 = 60$

PRACTICE

Use your calculator to solve these problems.

1. $\frac{4}{5} \times 500 =$
2. $\frac{1}{3} \times 726 =$
3. $\frac{2}{3} \times 1{,}500 =$
4. $\frac{5}{8} \times 312 =$
5. $\frac{5}{6} \times 84 =$
6. $\frac{4}{9} \times 234 =$
7. $\frac{3}{5} \times 600 =$
8. $\frac{4}{7} \times 1400 =$
9. $\frac{5}{9} \times 270 =$
10. $\frac{2}{3} \times 330 =$
11. $\frac{1}{2} \times 900 =$
12. $\frac{5}{8} \times 6400 =$
13. $\frac{3}{4} \times 1600 =$
14. $\frac{5}{6} \times 1800 =$
15. $\frac{7}{10} \times 1200 =$

CHAPTER REVIEW

CHAPTER SUMMARY

■ × and ÷	A fraction may be multiplied or divided by another fraction.	$\frac{3}{4} \times \frac{1}{2} =$
	A fraction may be multiplied or divided by a mixed number.	$\frac{3}{4} \div 2\frac{1}{2} =$
	A fraction may be multiplied or divided by a whole number.	$\frac{3}{4} \times 2 =$
■ Mixed Numbers	Before multiplying or dividing a mixed number, the mixed number must be changed to an improper fraction.	$2\frac{1}{4} = \frac{9}{4}$
■ Whole Numbers	Before multiplying or dividing a whole number, the whole number must be changed to an improper fraction. Make the whole number the numerator. Make the denominator 1.	$3 = \frac{3}{1}$
■ Inverting Fractions	Before dividing you must invert the second fraction in the problem. Make the numerator the denominator. Make the denominator the numerator.	$\frac{2}{3}$ becomes $\frac{3}{2}$

REVIEWING VOCABULARY

Use the correct words from the box to complete the sentences. Write on a separate sheet of paper.

mixed number	cancel	invert	proper fraction

1. When you ________ fractions, you make it easier to multiply or divide.

2. You must ________ the second fraction in order to solve a division problem.

3. A ________ is a number that is made up of a whole number and a fraction.

4. A ________ is a fraction whose numerator is smaller than its denominator.

Chapter Review

Chapter Quiz

A. Solve the problems below. Reduce the answers to lowest terms. Change any answer that is an improper fraction to a mixed number.

1. $\frac{2}{3} \times \frac{1}{2} =$

2. $\frac{2}{3} \div \frac{1}{2} =$

3. $\frac{3}{4} \div 4\frac{1}{5} =$

4. $4\frac{1}{5} \div \frac{3}{4} =$

5. $6\frac{2}{7} \div 2\frac{1}{5} =$

6. $2\frac{1}{5} \div 6\frac{2}{7} =$

7. $3\frac{3}{4} \div 2 =$

8. $2 \div 3\frac{3}{4} =$

9. $2 \times \frac{6}{7} =$

10. $2\frac{7}{10} \times 9 =$

11. $\frac{5}{18} \div \frac{1}{3} =$

12. $3\frac{1}{8} \times \frac{5}{8} =$

13. $\frac{6}{10} \div \frac{4}{5} =$

14. $\frac{8}{9} \div 1\frac{5}{6} =$

15. $\frac{7}{5} \div \frac{3}{8} =$

16. $\frac{4}{9} \times \frac{9}{10} =$

17. $7\frac{1}{2} \div \frac{2}{5} =$

18. $\frac{2}{13} \times \frac{7}{10} =$

19. $7 \times 8\frac{1}{3} =$

20. $4\frac{2}{12} \times 5\frac{5}{6} =$

1. $\frac{4}{6} \div \frac{2}{3} =$

2. $\frac{7}{8} \times \frac{1}{2} =$

3. $4\frac{5}{7} \times 6\frac{2}{8} =$

4. $7\frac{1}{9} \div 3\frac{2}{3} =$

5. $9\frac{1}{2} \div \frac{3}{4} =$

6. $\frac{15}{16} \div 2 =$

7. $5 \div \frac{4}{9} =$

8. $6\frac{7}{8} \div 4 =$

9. $8 \div 5\frac{2}{5} =$

10. $\frac{4}{15} \times 3\frac{1}{6} =$

B. Solve the problems below. Reduce answers to lowest terms. Change any answer that is an improper fraction to a mixed number.

1. $\frac{4}{6} \div \frac{2}{3} =$

2. $\frac{7}{8} \times \frac{1}{2} =$

3. $4\frac{5}{7} \times 6\frac{2}{8} =$

4. $7\frac{1}{9} \div 3\frac{2}{3} =$

5. $9\frac{1}{2} \div \frac{3}{4} =$

6. $\frac{15}{16} \div 2 =$

7. $5 \div \frac{4}{9} =$

8. $6\frac{7}{8} \div 4 =$

9. $8 \div 5\frac{2}{5} =$

10. $\frac{4}{15} \times 3\frac{1}{6} =$

C. Solve these word problems on a separate sheet of paper.

1. One box of books weighed $6\frac{1}{5}$ pounds. The second box weighed $\frac{3}{4}$ as much. How much does the second box weigh?

2. A piece of cloth is $3\frac{3}{4}$ yards long. Meg wants to use the cloth to make blouses. Each blouse takes up $1\frac{1}{4}$ yards. How many blouses can Meg make with the cloth she has?

3. Joshua can bench press 130 pounds. Henry can bench press $\frac{4}{5}$ as much. How much can Henry bench press?

4. Linda is making chocolate chip cookies. The recipe calls for $\frac{3}{4}$ cup of chocolate chips. How many cups of chocolate chips will she need for 3 recipes?

ADDING AND SUBTRACTING FRACTIONS

9

Skilled workers, such as this draftswoman, often need to work with fractions on the job. Can you think of other professions where workers need a thorough knowledge of fractions to do their jobs well?

Chapter Learning Objectives

1. Add fractions
2. Subtract fractions
3. Rename mixed numbers
4. Add mixed numbers
5. Subtract mixed numbers
6. Solve problems with fractions and mixed numbers

WORDS TO KNOW

- **like mixed numbers** mixed numbers whose fractions have the same denominator
- **unlike mixed numbers** mixed numbers whose fractions have different denominators

9.1 Adding Like Fractions

You have already learned that fractions that have the same denominator are called like fractions. Adding like fractions is easy. Follow the steps below.

Step 1: Add the numerators.

Step 2: Keep the same denominator.

Step 3: Reduce the fraction to the lowest terms.

$$\frac{3}{8} + \frac{3}{8} \qquad \frac{3}{8} + \frac{3}{8} = 6 \qquad \frac{3}{8} + \frac{3}{8} = \frac{6}{8} = \frac{3}{4}$$

PRACTICE

Find the sums in the problems below. Work on a separate sheet of paper.

1. $\frac{1}{3} + \frac{1}{3}$
2. $\frac{4}{17} + \frac{1}{17}$
3. $\frac{3}{12} + \frac{7}{12}$
4. $\frac{2}{7} + \frac{4}{7}$
5. $\frac{2}{15} + \frac{3}{15}$
6. $\frac{9}{20} + \frac{1}{20}$
7. $\frac{5}{8} + \frac{1}{8}$
8. $\frac{2}{10} + \frac{6}{10}$
9. $\frac{10}{18} + \frac{4}{18}$
10. $\frac{4}{9} + \frac{2}{9}$
11. $\frac{1}{6} + \frac{3}{6}$
12. $\frac{5}{20} + \frac{7}{20}$

9.2 Subtracting Like Fractions

Here is the method for subtracting like fractions. Follow the steps below.

Step 1: Subtract the numerators.

Step 2: Keep the same denominator.

Step 3: Reduce the fraction to the lowest terms.

$$\frac{7}{8} - \frac{3}{8} \qquad \frac{7}{8} - \frac{3}{8} = \frac{4}{\ } \qquad \frac{7}{8} - \frac{3}{8} = \frac{4}{8} = \frac{1}{2}$$

PRACTICE

Find the differences in the problems below. Work on a separate sheet of paper.

1. $\frac{2}{3} - \frac{1}{3}$

2. $\frac{9}{16} - \frac{5}{16}$

3. $\frac{19}{25} - \frac{14}{25}$

4. $\frac{16}{19} - \frac{5}{19}$

5. $\frac{7}{10} - \frac{1}{10}$

6. $\frac{18}{20} - \frac{4}{20}$

7. $\frac{8}{10} - \frac{7}{10}$

8. $\frac{7}{15} - \frac{4}{15}$

9. $\frac{8}{9} - \frac{3}{9}$

10. $\frac{17}{32} - \frac{9}{32}$

11. $\frac{12}{15} - \frac{2}{15}$

12. $\frac{8}{22} - \frac{3}{22}$

13. $\frac{7}{15} - \frac{4}{15}$

14. $\frac{13}{18} - \frac{4}{18}$

15. $\frac{30}{35} - \frac{25}{35}$

16. $\frac{9}{21} - \frac{6}{21}$

9.3 Adding Unlike Fractions

You have already learned that fractions that have different denominators are called unlike fractions.

Before you can add unlike fractions, you need to change them to equivalent like fractions. Follow the steps below.

Step 1: Change the fractions so that they have a common denominator (the least common denominator).

Step 2: Add the numerators.

Step 3: Keep the common denominator.

Step 4: Reduce the fraction to the lowest terms if possible.

For help in Step 1, turn back to page 118.

$\frac{1}{3} + \frac{1}{4} =$

$$\frac{1}{3} = \frac{4}{12}$$
$$+\frac{1}{4} = \frac{3}{12}$$
$$\frac{7}{12}$$

The least common denominator is the least common multiple (LCM) of both denominators.

PRACTICE

Find the sums of the problems below. Reduce the sums to the lowest terms. Work on a separate sheet of paper.

1. $\frac{2}{3} + \frac{2}{5}$

2. $\frac{7}{9} + \frac{1}{6}$

3. $\frac{3}{7} + \frac{2}{14}$

4. $\frac{9}{12} + \frac{6}{8}$

5. $\frac{3}{4} + \frac{1}{6}$

6. $\frac{1}{4} + \frac{3}{5}$

7. $\frac{4}{5} + \frac{2}{15}$

8. $\frac{17}{20} + \frac{3}{10}$

9. $\frac{2}{8} + \frac{1}{4}$

10. $\frac{5}{12} + \frac{3}{4}$

11. $\frac{7}{9} + \frac{6}{27}$

12. $\frac{5}{7} + \frac{11}{28}$

13. $\frac{5}{12} + \frac{2}{3}$

14. $\frac{2}{6} + \frac{1}{5}$

15. $\frac{7}{8} + \frac{2}{3}$

16. $\frac{3}{8} + \frac{19}{40}$

9.4 Subtracting Unlike Fractions

Here is the method for subtracting unlike fractions. Follow the steps below.

$\frac{4}{5} - \frac{1}{2} =$

Step 1: Change the fractions so that they have a common denominator (the least common denominator).

$\frac{4}{5} = \frac{8}{10}$

$-\frac{1}{2} = \frac{5}{10}$

$\frac{3}{10}$

Step 2: Subtract the numerators.

Step 3: Keep the common denominator.

Step 4: Reduce the fraction to the lowest terms.

PRACTICE

Find the differences in the problems below. Reduce the differences to the lowest terms. Work on a separate sheet of paper.

1. $\frac{1}{2} - \frac{1}{3}$

2. $\frac{3}{4} - \frac{1}{2}$

3. $\frac{6}{7} - \frac{2}{3}$

4. $\frac{7}{10} - \frac{1}{4}$

5. $\frac{2}{5} - \frac{1}{10}$

6. $\frac{7}{8} - \frac{3}{4}$

7. $\frac{5}{9} - \frac{2}{6}$

8. $\frac{11}{12} - \frac{4}{6}$

9. $\frac{7}{9} - \frac{2}{3}$

10. $\frac{5}{6} - \frac{3}{4}$

11. $\frac{7}{18} - \frac{2}{6}$

12. $\frac{13}{20} - \frac{2}{5}$

13. $\frac{3}{5} - \frac{4}{10}$

14. $\frac{12}{20} - \frac{4}{10}$

15. $\frac{5}{6} - \frac{1}{5}$

16. $\frac{13}{15} - \frac{7}{10}$

9.5 Renaming Mixed Numbers

You have already learned that a fraction expresses part of a whole unit. You can also express the whole unit as a fraction.

$$1 = \frac{1}{1} \quad 1 = \frac{2}{2} \quad 1 = \frac{3}{3} \quad 1 = \frac{4}{4} \quad 1 = \frac{5}{5} \ldots$$

You can use this idea to help you rename whole numbers as mixed numbers. For example:

Rename 9 as a mixed number. Follow the steps below.

Step 1: $9 = 8 + 1$

Step 2: Choose a denominator. 3

Step 3: Change 1 to a fraction with that denominator.

$9 = 8\frac{3}{3}$

PRACTICE

A. Change each whole number to a mixed number. Use a different denominator in each mixed number.

1. 4 **2.** 6 **3.** 7 **4.** 19 **5.** 23

Sometimes you will need to rename mixed numbers. The method is simple. For example:

Rename $9\frac{3}{4}$. Follow the steps below.

Step 1: $9 = 8 + 1$

Step 2: Change 1 to a fraction with the denominator 4.

Step 3: $\frac{3}{4} + \frac{4}{4} = \frac{7}{4}$ $\quad 9\frac{3}{4} = 8\frac{7}{4}$

PRACTICE

B. Rename each mixed number below. Write your answers on a separate sheet of paper.

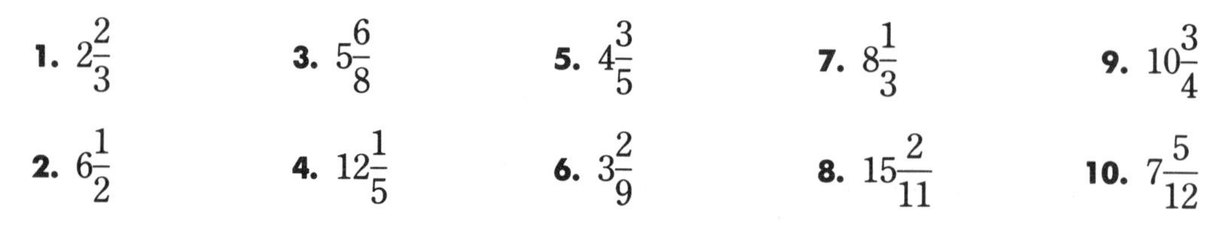

1. $2\frac{2}{3}$ **3.** $5\frac{6}{8}$ **5.** $4\frac{3}{5}$ **7.** $8\frac{1}{3}$ **9.** $10\frac{3}{4}$

2. $6\frac{1}{2}$ **4.** $12\frac{1}{5}$ **6.** $3\frac{2}{9}$ **8.** $15\frac{2}{11}$ **10.** $7\frac{5}{12}$

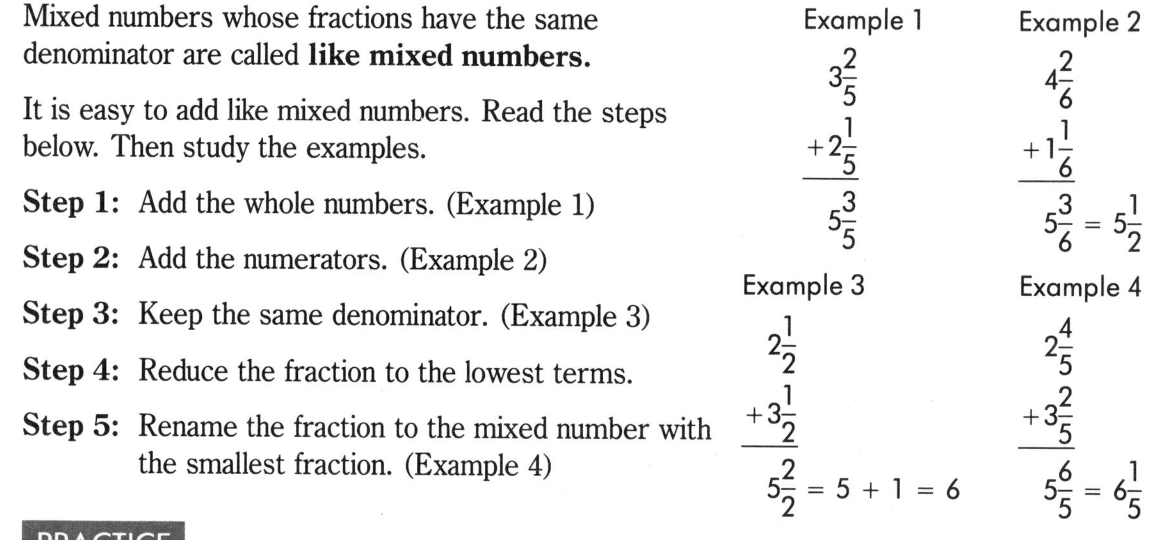

9.6 Adding Like Mixed Numbers

Mixed numbers whose fractions have the same denominator are called **like mixed numbers.**

It is easy to add like mixed numbers. Read the steps below. Then study the examples.

Step 1: Add the whole numbers. (Example 1)

Step 2: Add the numerators. (Example 2)

Step 3: Keep the same denominator. (Example 3)

Step 4: Reduce the fraction to the lowest terms.

Step 5: Rename the fraction to the mixed number with the smallest fraction. (Example 4)

Example 1

$$3\frac{2}{5} + 2\frac{1}{5} = 5\frac{3}{5}$$

Example 2

$$4\frac{2}{6} + 1\frac{1}{6} = 5\frac{3}{6} = 5\frac{1}{2}$$

Example 3

$$2\frac{1}{2} + 3\frac{1}{2} = 5\frac{2}{2} = 5 + 1 = 6$$

Example 4

$$2\frac{4}{5} + 3\frac{2}{5} = 5\frac{6}{5} = 6\frac{1}{5}$$

PRACTICE

Find the sums in the problems below. Work on a separate sheet of paper.

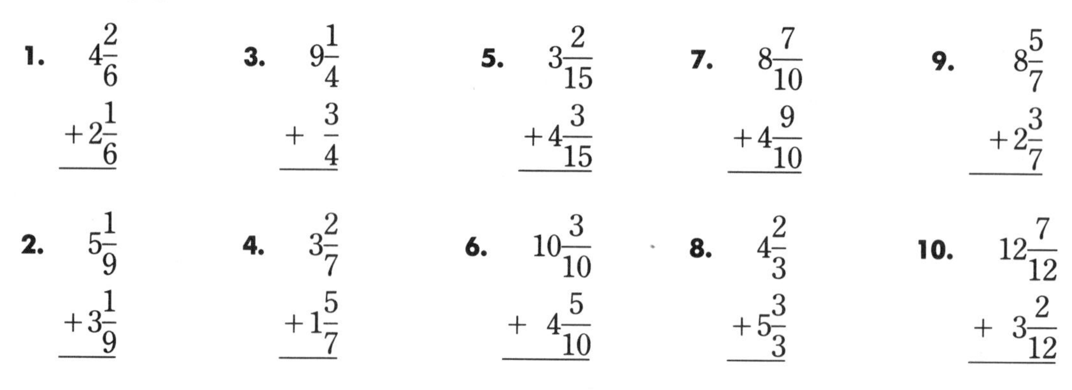

1. $4\frac{2}{6} + 2\frac{1}{6}$ **3.** $9\frac{1}{4} + \frac{3}{4}$ **5.** $3\frac{2}{15} + 4\frac{3}{15}$ **7.** $8\frac{7}{10} + 4\frac{9}{10}$ **9.** $8\frac{5}{7} + 2\frac{3}{7}$

2. $5\frac{1}{9} + 3\frac{1}{9}$ **4.** $3\frac{2}{7} + 1\frac{5}{7}$ **6.** $10\frac{3}{10} + 4\frac{5}{10}$ **8.** $4\frac{2}{3} + 5\frac{3}{3}$ **10.** $12\frac{7}{12} + 3\frac{2}{12}$

9.7 Subtracting Like Mixed Numbers

Here is the method for subtracting liked mixed numbers. Read the steps below. Then study the example.

Step 1: Compare the two fractions. If the first (top) fraction is smaller, rename it so that it becomes larger.

Step 2: Subtract the fractions.

Step 3: Subtract the whole numbers.

Step 4: Reduce the fraction to the lowest terms.

Step 5: Rename the fraction to the mixed number with the smallest fraction.

$$\begin{array}{rcr} 9\frac{1}{5} & = & 8\frac{6}{5} \\ -2\frac{2}{5} & = & 2\frac{2}{5} \\ \hline & & 6\frac{4}{5} \end{array}$$

PRACTICE

Find the differences in the problems below. Work on a separate sheet of paper.

1. $4\frac{2}{3} - 1\frac{1}{3}$

2. $7\frac{7}{8} - 4\frac{3}{8}$

3. $9\frac{4}{5} - 6\frac{2}{5}$

4. $17\frac{9}{30} - 16\frac{3}{30}$

5. $24\frac{1}{10} - 10\frac{3}{10}$

6. $19\frac{2}{7} - 2\frac{5}{7}$

7. $13\frac{7}{20} - 4\frac{3}{20}$

8. $38\frac{5}{12} - 3\frac{7}{12}$

9. $29\frac{4}{10} - 24\frac{5}{10}$

10. $11\frac{5}{9} - 7\frac{2}{9}$

11. $15\frac{2}{3} - 5\frac{1}{3}$

12. $8\frac{3}{5} - 2\frac{6}{10}$

13. $9\frac{3}{4} - 6\frac{1}{4}$

14. $11\frac{7}{9} - 5\frac{8}{9}$

15. $16\frac{7}{11} - 9\frac{4}{11}$

9.8 Adding Unlike Mixed Numbers

Mixed numbers whose fractions have different denominators are called **unlike mixed numbers.**

To add unlike mixed numbers remember how to add unlike fractions. Read the steps below. Then study the example.

Step 1: Change the fractions so that they have a common denominator (the least common denominator).

Step 2: Add the fractions.

Step 3: Add the whole numbers.

Step 4: Reduce the fraction to the lowest terms.

Step 5: Rename the fraction to the mixed number with the smallest fraction.

$$\begin{array}{rcr} 3\frac{1}{8} & = & 3\frac{1}{8} \\ +2\frac{1}{4} & = & 2\frac{2}{8} \\ \hline & & 5\frac{3}{8} \end{array}$$

PRACTICE

Find the sums in the problems below. Reduce the fractions to lowest terms. Work on a separate sheet of paper.

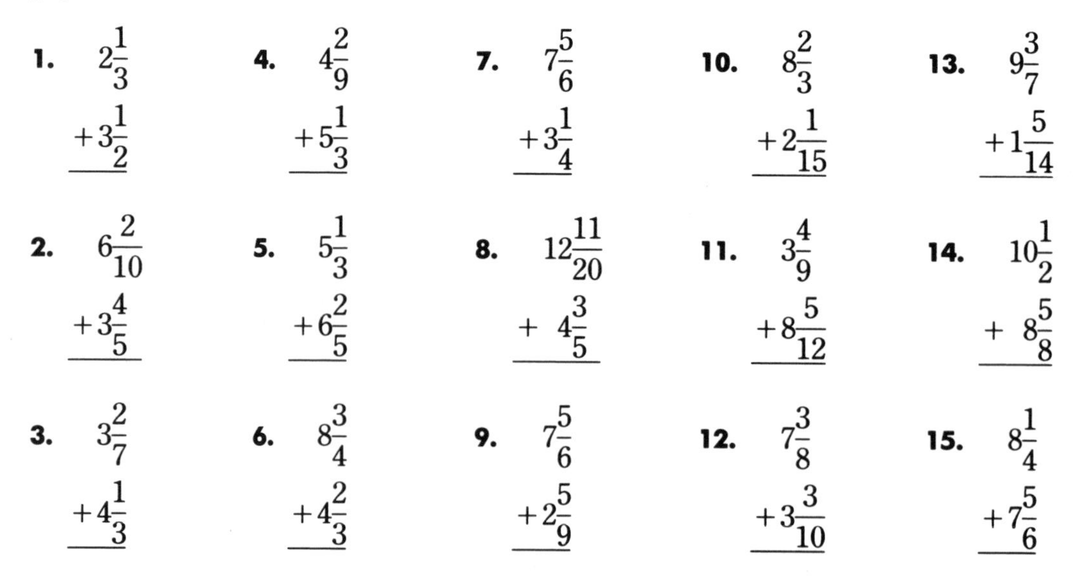

1. $2\frac{1}{3} + 3\frac{1}{2}$
2. $6\frac{2}{10} + 3\frac{4}{5}$
3. $3\frac{2}{7} + 4\frac{1}{3}$
4. $4\frac{2}{9} + 5\frac{1}{3}$
5. $5\frac{1}{3} + 6\frac{2}{5}$
6. $8\frac{3}{4} + 4\frac{2}{3}$
7. $7\frac{5}{6} + 3\frac{1}{4}$
8. $12\frac{11}{20} + 4\frac{3}{5}$
9. $7\frac{5}{6} + 2\frac{5}{9}$
10. $8\frac{2}{3} + 2\frac{1}{15}$
11. $3\frac{4}{9} + 8\frac{5}{12}$
12. $7\frac{3}{8} + 3\frac{3}{10}$
13. $9\frac{3}{7} + 1\frac{5}{14}$
14. $10\frac{1}{2} + 8\frac{5}{8}$
15. $8\frac{1}{4} + 7\frac{5}{6}$

9.9 Subtracting Unlike Mixed Numbers

Here is the method for subtracting unlike mixed numbers. Study the steps and the example below.

Step 1: Change the fractions so that they have a common denominator (the least common denominator).

Step 2: Rename if necessary.

Step 3: Subtract the fractions.

Step 4: Subtract the whole numbers.

Step 5: Reduce the fraction to the lowest terms.

Step 6: Rename if possible.

$$\begin{array}{rcr} 6\frac{3}{4} & = & 6\frac{9}{12} \\ -2\frac{1}{3} & = & 2\frac{4}{12} \\ \hline & & 4\frac{5}{12} \end{array}$$

PRACTICE

Find the differences in the problems below. Reduce the fractions to lowest terms. Work on a separate sheet of paper.

1. $3\frac{1}{2} - 1\frac{1}{3}$

2. $8\frac{5}{6} - 2\frac{7}{18}$

3. $7\frac{4}{5} - 2\frac{2}{5}$

4. $15\frac{1}{2} - 3\frac{4}{5}$

5. $25\frac{6}{7} - 4\frac{1}{14}$

6. $18\frac{3}{9} - 5\frac{1}{3}$

7. $30\frac{1}{8} - 10\frac{2}{9}$

8. $18\frac{3}{6} - 9\frac{2}{10}$

9. $11\frac{5}{8} - 7\frac{1}{3}$

10. $22\frac{7}{9} - 8\frac{3}{4}$

11. $13\frac{5}{7} - 8\frac{2}{21}$

12. $14\frac{5}{6} - 9\frac{3}{4}$

13. $17\frac{2}{5} - 9\frac{7}{10}$

14. $10\frac{5}{9} - 7\frac{2}{3}$

15. $12\frac{2}{5} - 8\frac{5}{6}$

9.10 Solving Problems with Fractions

Solving word problems with fractions is easy. Just follow the steps below and look for clue words.

Steps in Solving Word Problems

Read this problem. Follow the steps to solve it.

Terry had $6\frac{7}{8}$ yards of green fabric. She also had $4\frac{3}{8}$ yards of blue fabric. How many more yards of green than blue fabric did she have?

Step 1: Read the problem.

Step 2: Learn what you must find out.

How many more yards of green fabric did Terry have?

Step 3: Notice the clue words. Clue words help you solve the problem. *How many more* tells you to find the difference between two numbers. To solve this problem, you should subtract.

Clue words: how many more

Step 4: Find the numbers. Which is the larger number? Which number will you subtract from the larger number?

Larger Number: $6\frac{7}{8}$

Number to Subtract: $4\frac{3}{8}$

Step 5: Write the problem. Solve it.

Step 6: Check the answer. Ask yourself if it makes sense.

$$\begin{array}{r} 6\frac{7}{8} \\ -\,4\frac{3}{8} \\ \hline 2\frac{4}{8} \end{array} = 2\frac{1}{2}$$

Clue Words

These words tell you to find the difference between numbers. That means you will need to subtract.

how many more *how many fewer* *how much more* *remain*
how much larger *how much shorter* *left* *difference*

These words tell you to put numbers together. That means you will need to add.

in all *together* *altogether* *total* *both*

Reminder: Not all problems have clue words.

PROBLEMS TO SOLVE

Carefully read each problem below. Decide whether to add or subtract. Then solve the problems on a separate sheet of paper.

1. Noreen started her trip with $15\frac{3}{4}$ gallons of gas in her car. When she finished, there were $2\frac{1}{4}$ gallons. How many gallons had she used?
2. Marla travels $8\frac{4}{5}$ miles to work each day. Don travels $6\frac{1}{2}$ miles. How many more miles a day does Marla travel?
3. Kathie bought $1\frac{1}{2}$ pounds of meat on Monday. On Tuesday she bought $2\frac{1}{2}$ pounds. How much meat did she buy altogether?
4. Tom worked $5\frac{1}{2}$ hours on Monday. He worked $6\frac{3}{4}$ hours on Wednesday. On Friday he worked $4\frac{1}{4}$ hours. How many hours did he work in all?
5. Elizabeth was knitting a sweater. She knitted $7\frac{3}{5}$ inches one day. The next day she knitted $5\frac{5}{6}$ inches. How many more inches did she knit the first day than the second?
6. Norm drove $7\frac{3}{8}$ miles in the morning. He drove $13\frac{2}{3}$ miles in the afternoon. How many miles did he drive that day?

MATHEMATICS IN YOUR LIFE: Hours of Work

Before each holiday season, Top Department Store hires part-time workers. Helen's job is to keep track of how many hours each person works each week. Here is Matthew Parker's work record for one week.

Employee	M	T	W	Th	F	S
Parker, Matthew	$3\frac{3}{4}$	0	$4\frac{1}{2}$	3	$5\frac{1}{3}$	3

In order to find out how many hours Matthew worked, Helen had to do these things:

1. Find the least common denominator.
2. Add the hours.
3. Reduce the sum to lowest terms.

$$\begin{array}{rcr} 3\frac{3}{4} & = & 3\frac{9}{12} \\ 0 & = & 0 \\ 4\frac{1}{2} & = & 4\frac{6}{12} \\ 3 & = & 3 \\ 5\frac{1}{3} & = & 5\frac{4}{12} \\ +3 & & 3 \\ \hline & & 18\frac{19}{12} = 19\frac{7}{12} \end{array}$$

PRACTICE

Copy the chart on the next page on a separate sheet of paper.

1. Find out how many hours each person worked. Use the steps above. Add the totals to your chart.
2. Who worked the most hours?
3. How many hours did they work all together?

Employee	M	T	W	Th	F	S	Total
Fedya, Nora	4	$3\frac{1}{2}$	0	5	5	6	
Jones, Martin	5	4	$2\frac{3}{4}$	$3\frac{1}{2}$	4	7	
Lopez, Antonio	$3\frac{3}{4}$	0	5	$4\frac{2}{3}$	5	6	
Chung, Corinne	6	3	$4\frac{1}{2}$	5	$3\frac{1}{4}$	7	

9.11 Using Your Calculator: A Game of Odd and Even

Here is a game you and a friend can play on a calculator. In this game each digit from 0 to 9 can be used only once. The function keys + − × and ÷ can be used more than once. Neither player can multiply or divide by 0. Before you begin, write the numbers from 0 to 9 on a sheet of paper. As you use each number draw a line through it.

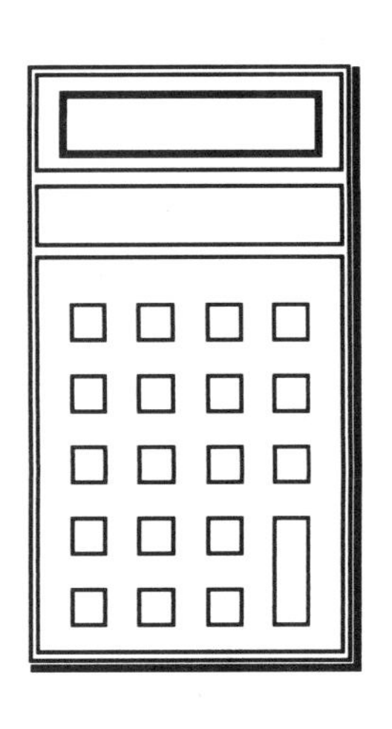

The object of the game is to choose either odd or even and make a number turn out that way. For example, Player Number 1 enters a number on the calculator and chooses either odd or even. Player Number 2 chooses a function (either + − × or ÷). Then Player Number 1 chooses another number. The object for Player Number 1 is to get the two numbers when used with the function to turn out the way he has chosen.

Players take turns choosing numbers and picking either odd or even. Play continues until all the numbers have been used.

CHAPTER REVIEW

CHAPTER SUMMARY

- **+ and −** A fraction may be added to or subtracted from another fraction.

  A mixed number may be added to or subtracted from another mixed number.

  You can only add or subtract fractions or mixed numbers that have common denominators.

- **Renaming** Sometimes you have to rename whole and mixed numbers in order to subtract or to make an answer simple.

- **Problem Solving** Clue words help you decide whether to add or subtract to solve a word problem with fractions.

REVIEWING VOCABULARY

Use the correct words from the box to complete the sentences. Write on a separate sheet of paper.

common	unlike mixed numbers	like mixed numbers	reduce

1. You must find the least common denominator before you can add or subtract ________ .
2. ________ have the same denominator.
3. To add or subtract unlike fractions you must first find a ________ denominator.
4. Always ________ fractions to their lowest terms.

Chapter Review

Chapter Quiz

A. Rename these mixed numbers on a separate sheet of paper. The renamed mixed number will have a smaller whole number and an improper fraction.

1. $17\frac{2}{3}$ **2.** $8\frac{5}{7}$ **3.** $9\frac{2}{5}$ **4.** $6\frac{1}{2}$ **5.** $81\frac{2}{15}$

B. Solve these problems on a separate sheet of paper. Reduce answers to lowest terms when possible.

1. $\frac{4}{5} - \frac{1}{2}$

2. $\frac{3}{4} + \frac{2}{6}$

3. $9\frac{7}{9} - 7\frac{5}{9}$

4. $\frac{7}{9} + \frac{2}{3}$

5. $\frac{7}{8} - \frac{1}{3}$

6. $8\frac{3}{4} + 9\frac{4}{16}$

7. $5\frac{4}{7} + 2\frac{2}{7}$

8. $6\frac{1}{8} + 5\frac{3}{8}$

9. $35\frac{3}{8} - 7\frac{3}{8}$

10. $13\frac{7}{8} - 7\frac{2}{3}$

11. $4\frac{2}{5} + 7\frac{2}{3}$

12. $12\frac{2}{3} - 6\frac{1}{4}$

13. $15\frac{1}{6} - 8\frac{2}{3}$

14. $10\frac{1}{6} - 8\frac{2}{3}$

15. $27\frac{2}{9} - 5\frac{3}{4}$

C. Solve the word problems below on a separate sheet of paper.

1. It is $32\frac{4}{5}$ miles from Kingston to Clemens Falls. It is $32\frac{1}{8}$ miles from Clemens Falls to Fosterville. Al drove from Kingston to Clemens Falls. Then he drove on to Fosterville. How many miles did he drive in all?

2. One cake recipe calls for $2\frac{1}{4}$ cups of flour. Another cake recipe calls for $2\frac{5}{7}$ cups of flour. How much less flour does the first recipe call for?

3. There were 15 yards of cloth on a roll. Jessie sold $3\frac{1}{3}$ yards. How many yards were left?

UNIT TWO REVIEW

Solve these problems on a separate sheet of paper.

1. Change each fraction to lowest terms.

a. $\frac{4}{8}$ b. $\frac{7}{21}$ c. $\frac{5}{45}$ d. $\frac{9}{27}$

2. Change each fraction to higher terms. Use the new denominators next to each fraction.

a. $\frac{2}{5}; \frac{}{25}$ b. $\frac{3}{8}; \frac{}{64}$ c. $\frac{1}{2}; \frac{}{36}$ d. $\frac{2}{3}; \frac{}{48}$

3. Change the improper fractions to mixed numbers.

a. $\frac{12}{7}$ b. $\frac{9}{5}$ c. $\frac{11}{3}$ d. $\frac{9}{2}$

4. Change these mixed numbers to improper fractions.

a. $5\frac{3}{4}$ b. $7\frac{2}{9}$ c. $6\frac{5}{8}$ d. $2\frac{5}{12}$

5. $\frac{2}{3} \times \frac{5}{7} =$

6. $\frac{4}{5} \times \frac{2}{13} =$

7. $2\frac{3}{8} - \frac{5}{6} =$

8. $\frac{3}{4} + \frac{5}{6} =$

9. $\frac{4}{5} \div \frac{2}{3} =$

10. $16\frac{7}{8} \div 5 =$

11. A piece of cloth is 16 yards long. It is cut into pieces each $\frac{3}{4}$ of a yard in length. How many small pieces can be cut from the one large piece?

12. Claudia is building a bookshelf. She needs $8\frac{2}{3}$ feet of wood for the sides and $15\frac{3}{8}$ feet for the shelves. How much wood does she need altogether?

13. Murray lost $3\frac{3}{8}$ pounds during the first week of his diet. He lost $5\frac{1}{3}$ pounds the second week. How much more weight did he lose the second week than the first?

UNIT THREE

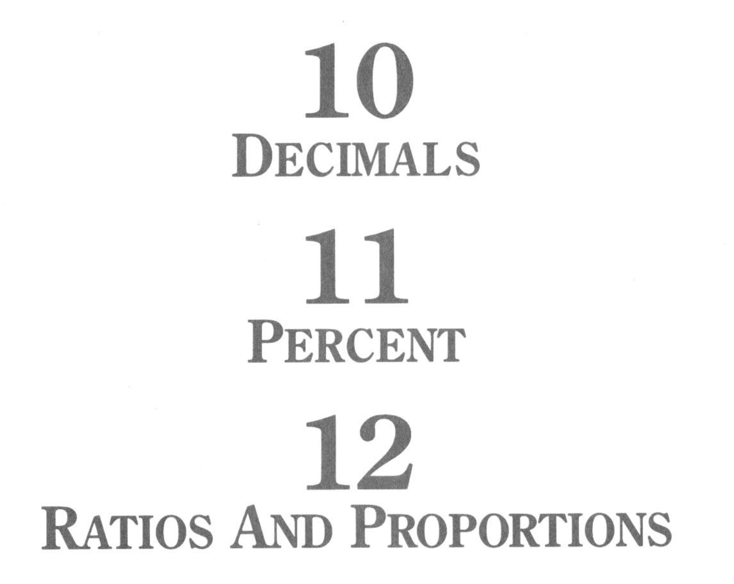

10
DECIMALS

11
PERCENT

12
RATIOS AND PROPORTIONS

DECIMALS

10

The chart behind this man tells him how much interest his money will earn with different savings certificates. Each of the numbers on the chart is a decimal. Can you think of anywhere else besides a bank where you often see decimals?

Chapter Learning Objectives

1. Understand the meaning of decimals
2. Explain and write decimals
3. Add decimals
4. Subtract decimals
5. Multiply decimals
6. Divide decimals
7. Change decimals to fractions
8. Solve problems with decimals

WORDS TO KNOW

- **decimal** a number written with a dot followed by places to the right; the digits to the right of the dot stand for less than one whole
- **decimal point** a period placed to the left of a decimal
- **mixed decimal** a number containing a whole number and a decimal number

10.1 What Is a Decimal?

A **decimal** is a number that is written with a dot followed by places to the right. Each of the digits to the right stands for less than a whole number. The dot is the **decimal point.**

Each of these decimals stands for less than a whole.

.1 .01 .38 .552 .1639

A decimal may also contain digits to the left of the decimal point. The digits to the left of the decimal point stand for a whole number. Numbers like these are called **mixed decimals.**

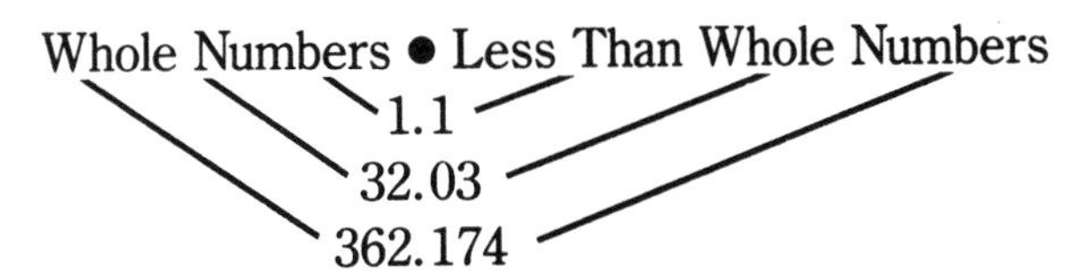

To rename a whole number as an equivalent decimal, write a decimal point to the right and add a zero. For example: 23 = 23.0

PRACTICE

Write the decimals below on a separate sheet of paper. Draw a line under all the digits that stand for less than one whole.

1. 45.067
2. 164.35
3. 544.002
4. 40.17
5. 829.1
6. 391.24
7. 98.3157
8. 403.405

10.2 Reading and Writing Decimals

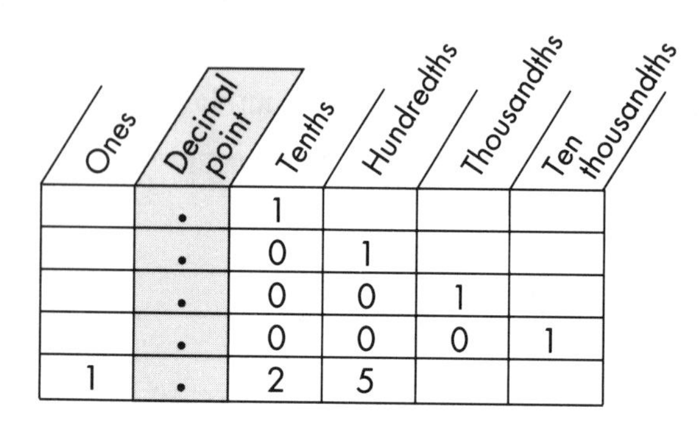

Ones	Decimal point	Tenths	Hundredths	Thousandths	Ten thousandths
	.	1			
	.	0	1		
	.	0	0	1	
	.	0	0	0	1
1	.	2	5		

Study how each decimal is read.
The decimal .1 is read one-tenth.
The decimal .01 is read one-one hundredth.
The decimal .001 is read one-one thousandth.
The decimal .0001 is read one-ten-thousandth.
The decimal 1.25 is read 1 and 25 hundredths.

PRACTICE

A. Read aloud the numbers on the chart below. Remember to read the decimal point as "and."

Thousands	Hundreds	Tens	Ones	Decimal point	Tenths	Hundredths	Thousandths	Ten thousandths
			9	.	4			
		1	2	.	7	2		
	1	5	6	.	8	9		
1	3	8	2	.	4	3	2	

B. On a separate sheet of paper, write the place *name* of the last digit in each decimal below. The first one has been done for you.

1. 32.7; tenths
2. 239.13
3. 62.472
4. 136.02
5. 500.19
6. 74.3009
7. 143.003
8. 25.6241

C. Rewrite the numbers below as decimals. Use a separate sheet of paper.

1. twenty-nine and four tenths
2. fifty-two and sixty-four hundredths
3. fifteen and two hundred fifty-two thousandths
4. fifty and three thousandths
5. twenty-one and thirty-two ten thousandths

10.3 Comparing Decimals

Study the pictures. Notice the different sizes of the parts in color.

Did You Know?
Numbers in the millions, billions, and trillions are often expressed as mixed decimals. For example, one of the most expensive paintings ever sold was purchased for 53.9 million dollars in 1987. It was called "Irises" by Vincent Van Gogh.

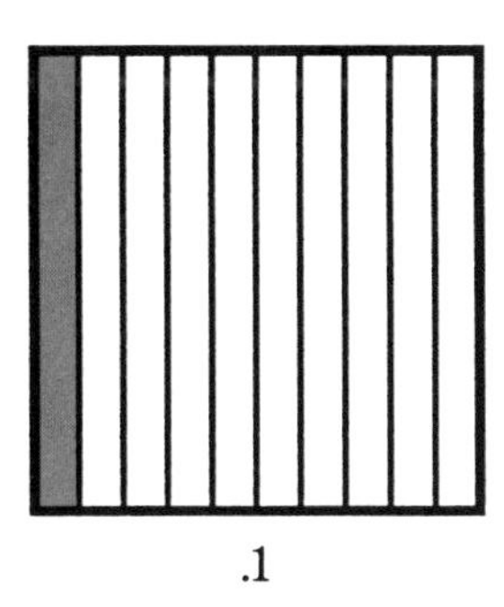

.1

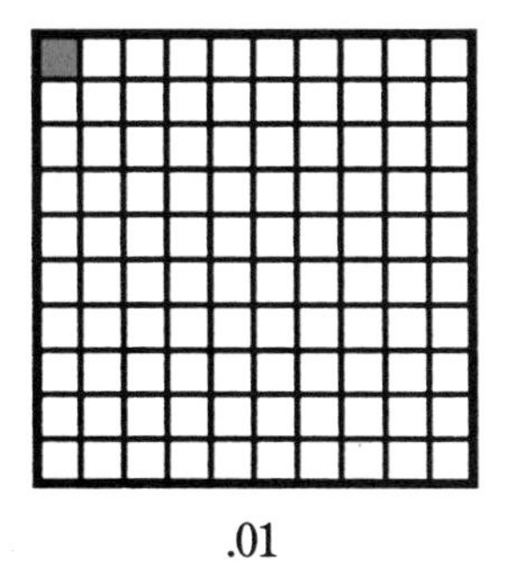

.01

The decimal .1 is larger than the decimal .01.
The decimal .01 is larger than the decimal .001.
The decimal .001 is larger than the decimal .0001.

The farther to the right a digit is placed in a decimal, the smaller the value of the number.

Study the pictures. Notice what part of each picture is in color.

The decimal .4 is larger than the decimal .3.
The decimal .3 is larger than the decimal .2, and so on.

The larger the number in a certain decimal place, the larger the decimal is.

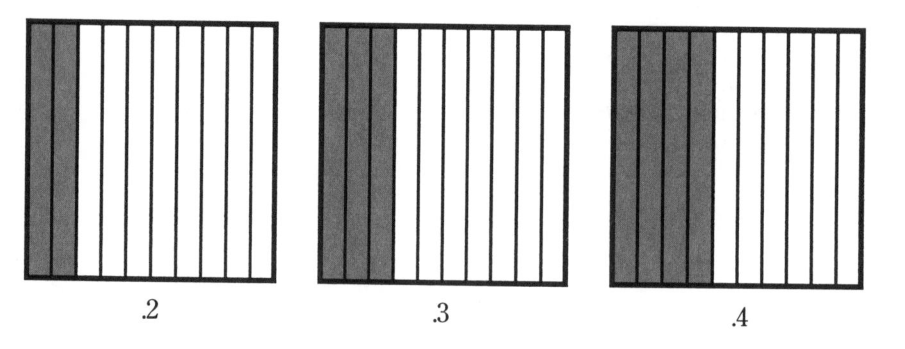

.2 .3 .4

Study the pictures. Notice what part of each picture is in color.

Writing or removing zeros from the end of a decimal does not change its value.

The same amount of each square is color.

.2 = .20

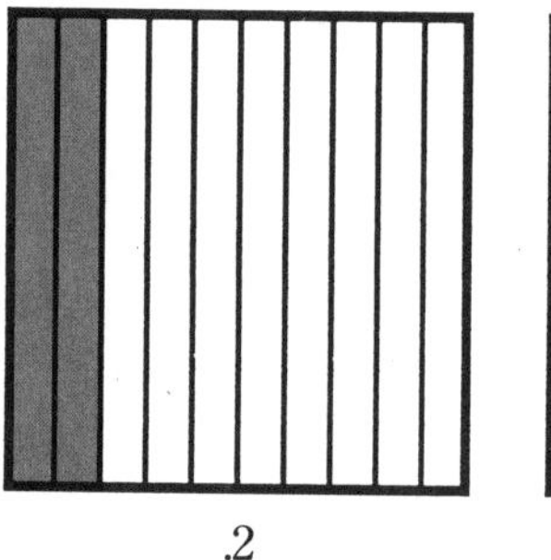

.2

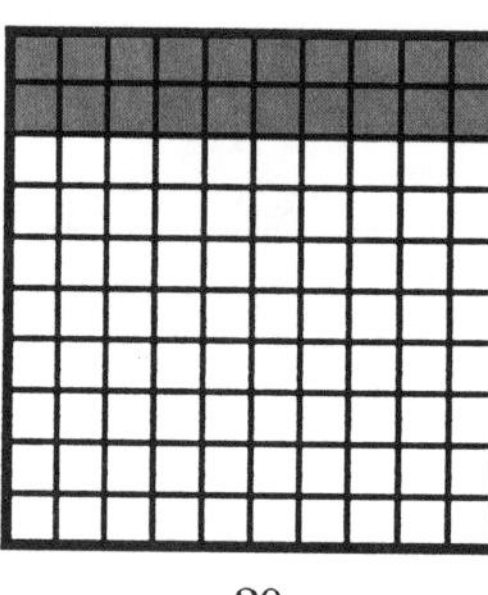

.20

To compare two decimals, follow the steps below. Then study the examples.

Step 1: Compare the digits to the left of the decimal point as two whole numbers. If one whole number is larger, then that decimal is larger.

Compare: 327.41 with 325.928

327 is larger than 325; therefore 327.41 is larger than 325.928.

Step 2: When the whole numbers are the same, compare the first place to the right of the decimal point. Write in zeros if necessary. If one digit is larger, then that decimal is larger.

Compare: 89.147 with 89.23

The whole numbers are the same (89 and 89). 2 is larger than 1; therefore 89.23 is larger than 89.147.

Step 3: Suppose the digits in the first place to the right of the decimal point are the same. Then compare the *next* place to the right. If one digit is larger, then that decimal is larger.
Compare: 1.924 with 1.93

The whole numbers are the same (1 and 1). The digits in the first place to the right of the decimal point are the same (9 and 9). 3 is larger than 2; therefore 1.93 is larger than 1.924.

PRACTICE

A. On a separate sheet of paper, write the larger decimal of each pair below.

1. .3 .03
2. .4 .45
3. .17 .20
4. .12 .21
5. .502 .52
6. .7 .4
7. .30 .91
8. .57 .06
9. .400 .500

B. Number a separate sheet of paper from 1 to 9. Next to each number, write the decimals in order of increasing size. The first one has been done for you.

1. .005 .5 .05
   .005 .05 .5
2. .025 .005 .3
3. .64 .650 .63
4. .72 .95 .53
5. .65 .32 .1
6. .700 .7 .72
7. .7 .2 .04
8. .5 .51 .49
9. .83 .8 .3

10.4 Adding Decimals

You add decimals the same way you add whole numbers. Follow the steps below. For example: Add 32.85 + .6 + 14. + 55.37.

Step 1: Write the decimal carefully.
Line up the decimal points one under the other.

Step 2: Write ending zeros so all decimals have the same number of places.

Step 3: Add the decimals just the way you add whole numbers.

Step 4: Put a decimal point in the sum. This decimal point should line up under the other decimal points.

$$\begin{array}{r} 32.85 \\ .60 \\ 14.00 \\ \underline{55.37} \\ 102.82 \end{array}$$

PRACTICE

Find the sums of the problems below. Do your work on a separate sheet of paper.

1. $\begin{array}{r} 304.25 \\ 89.00 \\ \underline{+426.70} \end{array}$

2. $\begin{array}{r} 159.80 \\ 341.80 \\ .5679 \\ \underline{+464.9000} \end{array}$

3. $\begin{array}{r} 59.02 \\ 162.91 \\ \underline{+494.90} \end{array}$

4. $\begin{array}{r} 403.50 \\ .93 \\ 4.10 \\ \underline{+\ \ 5.64} \end{array}$

5. $\begin{array}{r} 401.293 \\ 27.000 \\ \underline{+\ \ 65.005} \end{array}$

6. $\begin{array}{r} 290.04 \\ 3.50 \\ 381.00 \\ \underline{+506.00} \end{array}$

10.5 Subtracting Decimals

You subtract decimals the same way you subtract whole numbers. Follow the steps below. For example: Subtract 29.34 – 16.52; subtract 87 – 35.83

Step 1: Write the decimal carefully.
Line up the decimal points one under the other.

$$\begin{array}{r} 29.34 \\ -16.52 \\ \hline 12.82 \end{array}$$

Step 2: Write ending zeros so all decimals have the same number of decimal places.

$$\begin{array}{r} 87. \\ -35.83 \\ \hline \end{array}$$

Step 3: Subtract the decimals just the way you subtract whole numbers.

$$\begin{array}{r} 87.00 \\ -35.83 \\ \hline 51.17 \end{array}$$

Step 4: Put a decimal point in the difference. This decimal point should line up under the other decimal points.

PRACTICE

Find the differences in the problems below. Do your work on a separate sheet of paper.

1. $\begin{array}{r} 963.857 \\ -241.534 \\ \hline \end{array}$

2. $\begin{array}{r} 835.091 \\ -482.320 \\ \hline \end{array}$

3. $\begin{array}{r} 467.09 \\ -180.50 \\ \hline \end{array}$

4. $\begin{array}{r} 746.43 \\ -597.50 \\ \hline \end{array}$

5. $\begin{array}{r} 45.92 \\ -23.07 \\ \hline \end{array}$

6. $\begin{array}{r} 630.47 \\ -\ 45.28 \\ \hline \end{array}$

7. $\begin{array}{r} 529.870 \\ -\ 48.325 \\ \hline \end{array}$

8. $\begin{array}{r} 692.09 \\ -\ 98.20 \\ \hline \end{array}$

9. $\begin{array}{r} 410.34 \\ -283.16 \\ \hline \end{array}$

10. $\begin{array}{r} 835.10 \\ -594.20 \\ \hline \end{array}$

11. $\begin{array}{r} 89.654 \\ -26.000 \\ \hline \end{array}$

12. $\begin{array}{r} 382.00 \\ -\ 46.20 \\ \hline \end{array}$

13. $\begin{array}{r} 358.23 \\ -\ 69.95 \\ \hline \end{array}$

14. $\begin{array}{r} 298.77 \\ -169.84 \\ \hline \end{array}$

15. $\begin{array}{r} 385.77 \\ -156.29 \\ \hline \end{array}$

10.6 Multiplying Decimals

Here is the method for multiplying decimals. Follow the steps below.

Step 1: Line up the numbers one under the other.

Step 2: Multiply the decimals just the way you multiply whole numbers.

```
  15.34
×  4.2
  3068
 6136
 64.428
```

Step 3: Count the number of decimal places in the numbers you have multiplied. The whole numbers have 0 decimal places.

```
  15.34     2 decimal places
×  4.2     +1 decimal place

 64.428     3 decimal places
```

Step 4: Start at the right. Count to the left the same number of decimal places in the product. Put in a decimal point.

PRACTICE

A. Find the products in the problems below. Do your work on a separate sheet of paper.

1. 359.65×48.12
2. 471.03×246.21
3. 725.472×407.58
4. 683.01×89.5
5. 815.08×92.34
6. 627.45×380.69
7. 890.03×83.62
8. 721.549×94.08
9. 509.54×38.67
10. 947.003×263.408
11. 587.35×364.29
12. 843.207×258.396
13. 37.25×2.8
14. 622.5×5.02
15. 365.77×2.26

Multiplying Decimals by 10, 100, 1,000

It is easy to multiply decimals by 10, 100, or 1,000.
Follow the steps below.

Step 1: Count the zeros.

Step 2: Move the decimal point one place to the right for each zero. Fill in with more zeros if necessary.

32.3 × 10 = 32.3 = 323.0   32.3 × 100 = 32.30 = 3,230.0

32.3 × 1,000 = 32.300 = 32,300.0

4.93 × 10 = 4.93 = 49.3   4.93 × 100 = 4.93 = 493.0

4.93 × 1,000 = 4.930 = 4,930.0

.171 × 10 = .171 = 1.71   .171 × 100 = .171 = 17.1

.171 × 1,000 = .171 = 171.0

B. Find the products in the problems below. Do your work on a separate sheet of paper.

1. .2 × 10 =
2. .35 × 10 =
3. .05 × 100 =
4. 2.97 × 100 =
5. 45.06 × 100 =
6. .539 × 100 =
7. 42.4 × 100 =
8. 22.8 × 1,000 =
9. 72.4 × 100 =
10. .457 × 1,000 =
11. 86.73 × 1,000 =
12. .9 × 1,000 =
13. 3.38 × 10 =
14. 72.8 × 1,000 =
15. 45.602 × 100 =
16. 463.8 × 100 =
17. 50.22 × 100 =
18. .07 × 1,000 =
19. .005 × 100 =
20. 7.5 × 10 =
21. 48.05 × 100 =

10.7 Dividing Decimals

Dividing Decimals by Whole Numbers

Here is the method for dividing decimals by whole numbers. Follow the steps below. For example: Divide 93.8 by 35.

$35\overline{)93.8}$

Step 1: Find the decimal point in the dividend. Put the decimal point in the quotient directly above it.

$$\begin{array}{r} \downarrow \\ . \\ 35\overline{)93.8} \end{array}$$

Step 2: Divide just the way you would with whole numbers until all digits of the dividend have been used.

Step 3: Divide further by writing more ending zeros until there is no remainder.

$$\begin{array}{r} 2.68 \\ 35\overline{)93.80} \\ \underline{70} \\ 23\ 8 \\ \underline{21\ 0} \\ 2\ 80 \\ \underline{2\ 80} \end{array}$$

Sometimes a division problem will produce a quotient with a repeating pattern. When this happens, the division will not end itself. For example: Divide 1.4 by 33.

The pattern 42 is repeated, so division will not end itself. Draw a bar over the repeating part of the answer.

$$\begin{array}{r} .04\overline{242} \\ 33\overline{)1.40000} \\ \underline{1\ 32} \\ 80 \\ \underline{66} \\ 140 \\ \underline{132} \\ 80 \\ \underline{66} \\ 14 \end{array}$$

> In all future division problems, all remainders should be expressed as *fractions*.

PRACTICE

A. Find the quotients in the problems below. Do your work on a separate sheet of paper. The first one has been done for you.

1. $3\overline{).27}$ = .09	**4.** $2\overline{).492}$	**7.** $27\overline{)85.32}$	**10.** $12\overline{).150}$
2. $5\overline{).45}$	**5.** $6\overline{)36.12}$	**8.** $34\overline{)356.32}$	**11.** $8\overline{).526}$
3. $8\overline{).56}$	**6.** $12\overline{)74.76}$	**9.** $64\overline{)58.56}$	**12.** $76\overline{)12.54}$

Dividing Decimals by Decimals

Here is the method for dividing decimals by decimals. Follow the steps below.

Step 1: Make the divisor a whole number. To do this, move the decimal point to the right of the last digit.

$3.2\overline{)2.24}$

Step 2: Move the decimal point in the dividend the same number of places to the right.

Step 3: Put a decimal point in the quotient directly above the decimal point in the dividend.

Step 4: Divide the same way you would divide a decimal by a whole number.

```
      .7
 32)22.4
    22.4
       0
```

Sometimes it is necessary to write ending zeros to the dividend. You will need to produce the same number of places in the dividend as are in the divisor. For example:

To check your answer, multiply the quotient by the *original* divisor—the decimal.

```
         144
.125)18.000
     12 5
      5 50
      5 00
       500
       500
```

PRACTICE

A. Find the quotients in the problems below. Do your work on a separate sheet of paper.

1. $.4\overline{).8}$

2. $.15\overline{).45}$

3. $.12\overline{)1.56}$

4. $.07\overline{)4.921}$

5. $.6\overline{)72.18}$

6. $1.04\overline{)93.8}$

7. $.25\overline{)16.5}$

8. $.06\overline{)5.6}$

9. $6.8\overline{)44.2}$

10. $4.4\overline{)82.28}$

11. $.08\overline{)170.4}$

12. $3.8\overline{)62.7}$

Dividing Decimals by 10, 100, 1,000

It is easy to divide decimals by 10, 100, or 1,000. Follow the steps below.

Step 1: Count the zeros.

Step 2: Move the decimal point one place to the left for each zero. Add more zeros if necessary.

$32.3 \div 10 = 32.3 = 3.23$

$4.93 \div 10 = 4.93 = .493$

$.171 \div 10 = 0.171 = .0171$

$32.3 \div 100 = 32.3 = .323$

$4.93 \div 100 \div 04.93 = .0493$

$.171 \div 100 = 00.171 = .00171$

$32.3 \div 1{,}000 = 032.3 = .0323$

$4.93 \div 1{,}000 = 004.93 = .00493$

$.171 \div 1{,}000 = 000.171 = .000171$

B. Find the quotients in the problems below. Do your work on a separate sheet of paper.

1. $.62 \div 10 =$

2. $.3 \div 10 =$

3. $.08 \div 100 =$

4. $5.74 \div 100 =$

5. $12.09 \div 100 =$

6. $.834 \div 100 =$

7. $21.6 \div 100 =$

8. $38.6 \div 1{,}000 =$

9. $84.7 \div 100 =$

10. $.239 \div 1{,}000 =$

11. $58.09 \div 1{,}000 =$

12. $.4 \div 1{,}000 =$

13. $38 \div 10 =$

14. $74.5 \div 1{,}000 =$

15. $2.98 \div 10 =$

16. $360 \div 100 =$

17. $2.95 \div 100 =$

18. $.5 \div 1{,}000 =$

19. $.18 \div 100 =$

20. $7.23 \div 10 =$

10.8 Changing Decimals to Fractions

Here is the method for changing decimals to fractions. Follow the steps below. For example: Write .35 as a fraction.

Step 1: Use the digits of the decimal as the numerator. $.35 = 35$

Step 2: Look at the last place in the decimal. The value of this place is the denominator. $.35 =$ hundredths $\frac{35}{100}$

Step 3: Reduce the fraction to lowest terms. $.35 = 35 \text{ hundredths} = \frac{35}{100} = \frac{7}{20}$

$.003 = 3 \text{ thousandths} = \frac{3}{1000}$

PRACTICE

Change each decimal below to a fraction. Reduce the fraction to lowest terms when possible. Do your work on a separate piece of paper.

Do you remember how to reduce a fraction to lowest terms? Just divide the numerator and denominator by their greatest common factor.

1. .5
2. .05
3. .005
4. .0005
5. .76
6. .845
7. .682
8. .98
9. .328
10. .455
11. .8739
12. .0605
13. .4
14. .40
15. .32
16. .8
17. .375
18. .75
19. .10
20. .28

10.9 Estimation: Rounding Decimals

Rounding decimals can help you estimate answers.

The method for rounding decimals is almost the same as for rounding whole numbers. You can round decimals to the nearest tenth, hundredth, or to any place you choose. Follow the steps below.

		Nearest Tenth	Nearest Hundredth
Step 1:	Find the place you are rounding to.	2.56 (5 = tenth)	3.482 (8 = hundredth)
Step 2:	Look at the digit to the right. If that digit is 5 or more, add 1 to the place you are rounding to. If that digit is less than 5, make no changes in the place you are rounding to.	2.56 (6 = 5 or more)	3.482 (2 = less than 5)
Step 3:	Drop all digits to the right of the place you are rounding to. When working with money, always keep two decimal places, writing in zeros when necessary.	2.6	3.48

PRACTICE

Round each decimal below twice. First, round each decimal to the nearest tenth. Then, round each decimal to the nearest hundredth. Do your work on a separate sheet of paper.

1. .3759
2. 1.49628
3. 48.9948
4. 2.8724
5. 23.0597
6. 105.5296
7. 832.2356
8. 54.3590
9. 258.94603
10. 67.3005
11. 999.3219
12. 32.4067

10.10 Solving Problems with Decimals

You have already learned the steps to solve problems with whole numbers. You can use those same steps to solve problems with decimals. Read the problem on the next page. Follow the steps to solve it.

Darrell's pay for two weeks is $624.32. He spent $229.55 the first week. He spent $234.92 the second week. How much money did he have left?

Step 1: Read the problem.

Step 2: Learn what you must find out. In this problem you must find out two things.

How much money did Darrell spend in all? How much money was left?

Step 3: Notice the clue words. *Left* tells you that you will have to subtract after you find out how much Darrell spent in all.

Clue Word: left

Step 4: Notice the numbers.

$624.32
$229.55
$234.92

Step 5: Write the problems. Solve them.

$$\begin{array}{r} \$229.55 \\ +\$234.92 \\ \hline \$464.47 \end{array} \qquad \begin{array}{r} \$624.32 \\ -\$464.47 \\ \hline \$159.85 \end{array}$$

Step 6: Check the answers. Ask yourself if they make sense.

PROBLEMS TO SOLVE

Solve the problems below on a separate sheet of paper.

1. Jim walks to work and back every day. He walks 2.9 miles round trip each day. He works 5 days a week. How many miles does he walk each week? He works 49 weeks a year. How many miles does he walk in a year?
2. Margaret bought 3 cans of tomatoes. The tomatoes were $.43 a can. She bought 3 pounds of potatoes. The potatoes were $.27 a pound. How much change from $5.00 did she receive?
3. Pablo had $198.22 in his checking account. He put in $36. Then he wrote a check for $52.75. How much did he have left in his checking account?

MATHEMATICS IN YOUR LIFE: Keeping Track of Prices

Stan and Liz had a long shopping list. They wanted to keep track of how much they were spending. So they rounded off the prices to the nearest dime. Then they added these rounded numbers as they shopped. Here is part of their list.

Prices	Prices Rounded to the Nearest Dime
$ 5.75	$ 5.80
2.42	2.40
16.03	16.00

PRACTICE

Round the prices on each list below to the nearest dime. Add the rounded prices. Then add the actual prices.

1.	**2.**	**3.**	**4.**
$ 4.58	$ 3.21	$ 2.94	$ 17.32
8.02	64.49	6.58	8.21
6.84	100.33	109.49	7.62
2.28	5.28	8.45	.34
	2.98		.95

5. Stan and Liz have saved $150.00. They want to buy a chair for $79.95, a picture frame for $29.50, and a lamp for $45.00. Round these prices quickly to see if they have saved enough for all three purchases.

6. Round the numbers and rewrite these newspaper headlines.
96 Musicians Go Out on Strike
45,012 Fans Pack the Ballpark
$99,990 Stolen in Bank Robbery

10.11 Using Your Calculator: Decimals

You can use your calculator to work with decimals. Follow the steps below.

Step 1: Press [8] [9] [5] [·] [3] [2]

Step 2: Press [−]

Step 3: Press [2] [5] [6] [·] [0] [1]

Step 4: Press [=]

Step 5: Your answer should be: 639.31

PRACTICE

Solve these problems on a calculator.

1. 4.803 + 156.7 =
2. 60 − 7.52
3. 25.3 + 18.95 =
4. 88 − 26.52 =
5. 75.3 − 1.808 =
6. 26.5 + 1.733 =
7. 6.2 − .05 =
8. 47.9 + 76 =
9. 25.7 + 456 =
10. 175.8 − 24 =
11. 475.88 + 23.3 =
12. 27.43 − 1.995 =
13. 543.64 ÷ 86.93 =
14. 980.29 × .00005 =
15. 42.11 × 175.11 =
16. 127.10 ÷ 40.04 =
17. 5932.03 ÷ 89.17 =
18. 76.4105 × 549.87 =
19. 153.10 ÷ 37.12 =
20. 727.15 × 66.66 =
21. 7559.482 ÷ 65.203 =
22. 476.0102 × 37.03 =
23. 12.1015 ÷ 4.04 =
24. 145.10 ÷ 6.25 =

CHAPTER REVIEW

CHAPTER SUMMARY

- **Decimal** — A decimal contains a decimal point. The digits to the right of the decimal point stand for a number that is less than one. Decimals may be added, subtracted, multiplied, and divided.
- **Fractions** — Decimals may be changed to fractions. Fractions may be changed to decimals.
- **Rounding Decimals** — Decimals may be rounded. Rounding decimals can help to estimate answers.

REVIEWING VOCABULARY

Use the correct word or words from the box to complete the sentences. Write on a separate sheet of paper.

add	decimal	mixed decimal	decimal point	right

1. A ________ is a number written with a dot followed by places to the right.
2. A ________ is the period placed to the left of a decimal.
3. A number containing a whole number and a decimal is called a ____________ .
4. To ________ a decimal, you must always line up the decimal points.
5. Drop all digits to the ________ of the place you are rounding to.

Chapter Review

CHAPTER QUIZ

A. Solve these problems on a separate sheet of paper.

1. $\begin{array}{r} 582.609 \\ 69.286 \\ +829.05 \\ \hline \end{array}$

2. $\begin{array}{r} 794.087 \\ 49.357 \\ +918.36 \\ \hline \end{array}$

3. $\begin{array}{r} 936.29 \\ -546.38 \\ \hline \end{array}$

4. $\begin{array}{r} 845.46 \\ -291.53 \\ \hline \end{array}$

5. $\begin{array}{r} 658.897 \\ \times \quad 100 \\ \hline \end{array}$

6. $\begin{array}{r} 1075.48 \\ \times\ 684.02 \\ \hline \end{array}$

7. $\begin{array}{r} 3261.45 \\ \times\ 376.05 \\ \hline \end{array}$

8. $\begin{array}{r} 839.806 \\ \times 392.167 \\ \hline \end{array}$

9. $10\overline{)3.569}$

10. $48\overline{)39.64}$

11. $35\overline{)23.5}$

12. $2.7\overline{)56.03}$

B. Change these decimals to fractions on a separate sheet of paper. Reduce the fractions to lowest terms.

1. .75 **2.** .33 **3.** .08 **4.** .218 **5.** .0983

C. Two of the three numbers in each group below are equal. Write the number in each group that is *not equal* to the other two. Do your work on a separate sheet of paper.

1. .03 .003 $\frac{3}{100}$

2. .4 .40 $\frac{3}{5}$

3. .14 .141 $\frac{7}{50}$

4. .5 .05 $\frac{1}{20}$

5. .25 2.5 $\frac{1}{4}$

6. .36 .32 $\frac{8}{25}$

D. Round each decimal below twice. First round to the nearest tenth. Then round to the nearest hundredth.

1. 3.655
2. .115
3. 2.477
4. 3.061
5. .995
6. 23.602
7. 48.844
8. .057
9. 1.949

PERCENT

11

This photograph from the 1930s shows farm workers at a dinner table. During that decade, nearly 25 percent of the U.S. population lived and worked on farms. By the late 1980s, that number had decreased to 2 percent.

Chapter Learning Objectives

1. Define percent
2. Find the percent of a whole number
3. Change decimals, fractions, and percents
4. Compare decimals, fractions, and percents
5. Find whole numbers when you know the percent
6. Find the percent one number is of another
7. Solve problems with percents

WORDS TO KNOW

- **percent** a part of a whole that has been divided into 100 equal parts

11.1 What Is Percent?

A **percent** is a part of a whole that has been divided into 100 equal parts. The sign for percent is %.

The whole is 100%. One part of 100 parts is 1%.

11.2 Examples of Percent

100% equals a whole.

Any percent less than 100% is less than one whole. For example, 50% is less than 100%. 50% is less than one whole.

Any percent more than 100% is more than one whole. For example, 150% is more than 100%. 150% is more than one whole.

= 150%

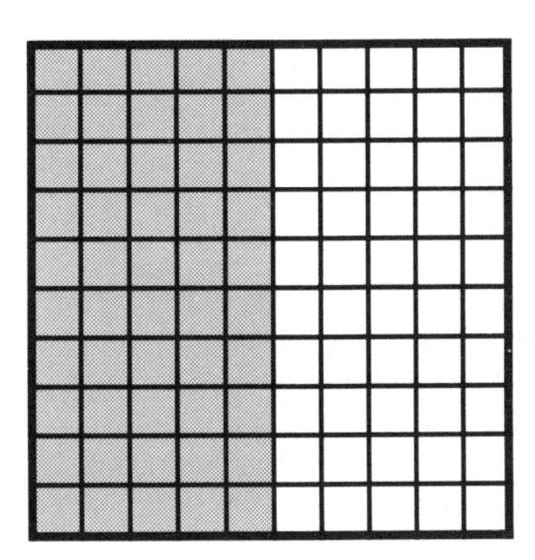

= 50%

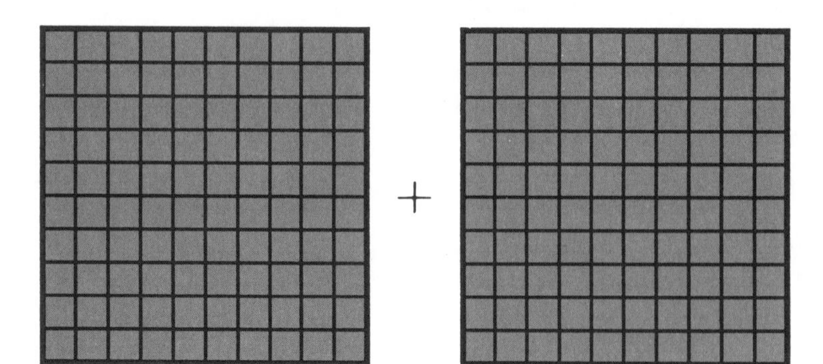

How many squares are in color?

200% equals two wholes.

250% equals more than two wholes.

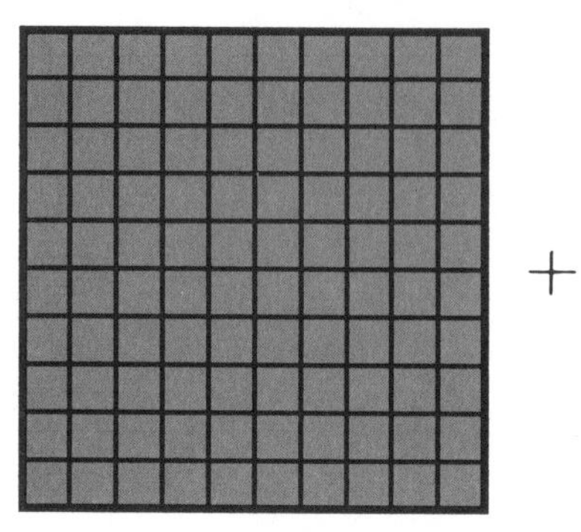

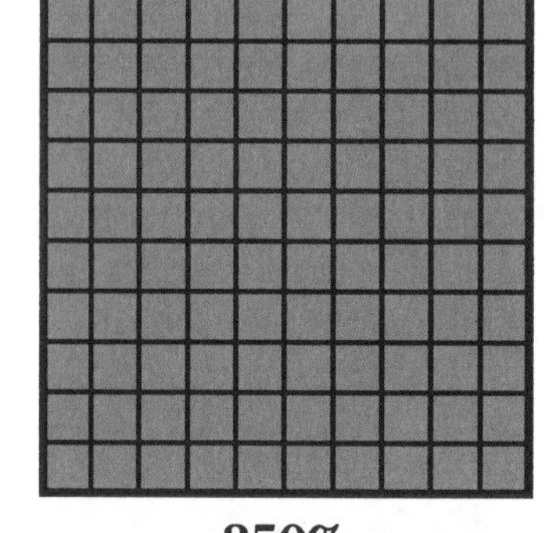

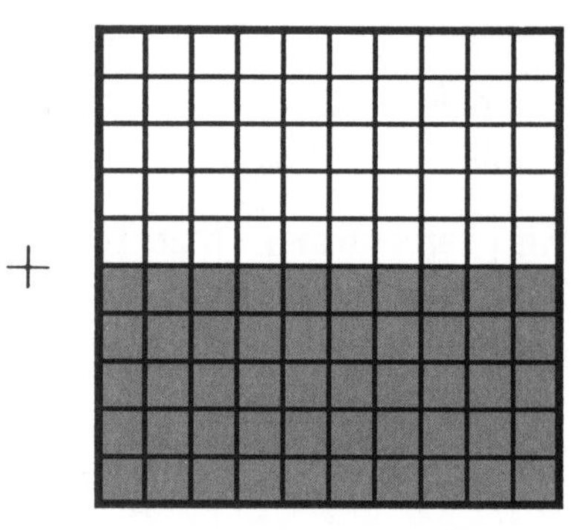

250%

How many squares are in color?

PRACTICE

On a separate sheet of paper write the percent of each of the following squares that is in color.

1.

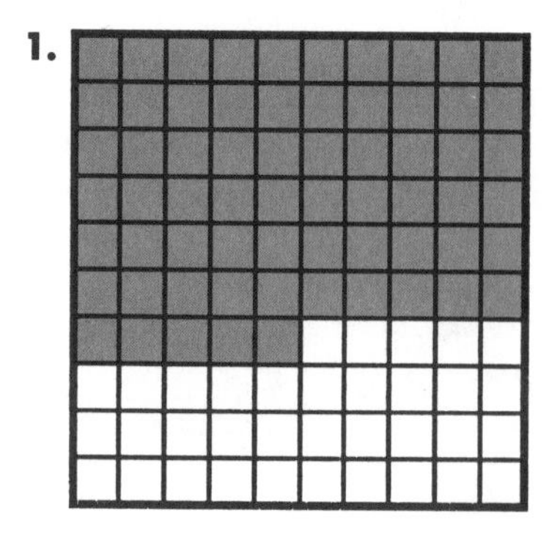

2.

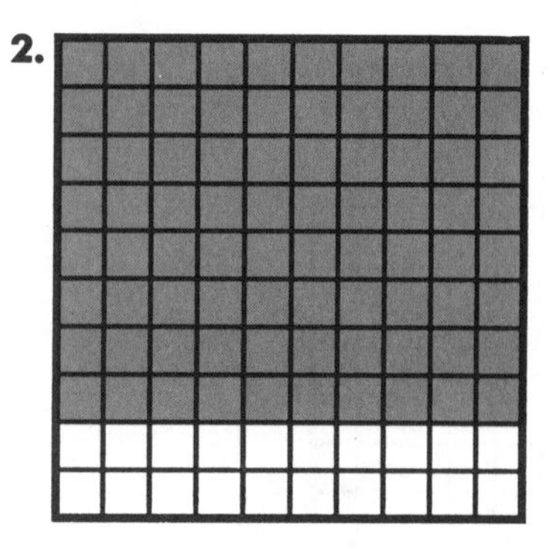

3.

4.

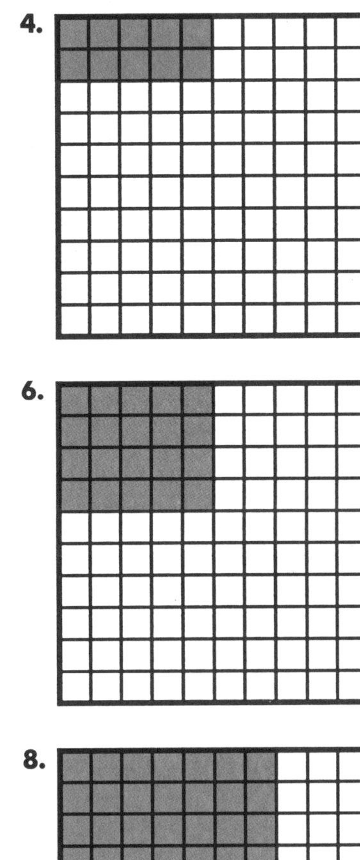

5.

6.

7.

8.

9.

10.

11.3 Changing Percents to Decimals

Equivalent percents, decimals, and fractions are different ways of talking about the same number value.

You cannot multiply or divide by a percent. You must first change the percent to an equivalent decimal or fraction.

It is easy to change percents into decimals or decimals into percents. Percents and decimals both refer to parts of a whole that has been divided into 100 equal parts.

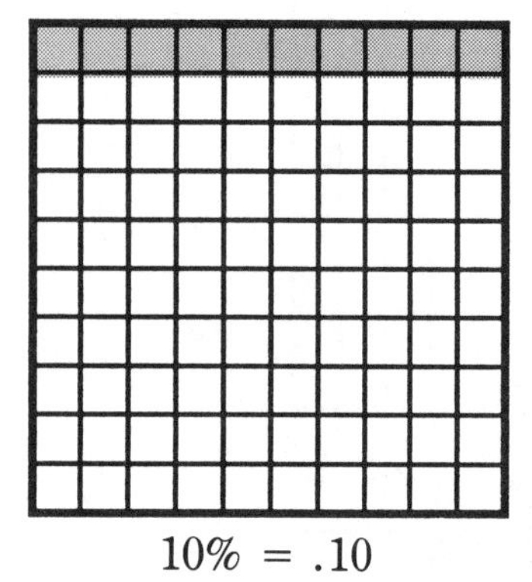

10% = .10

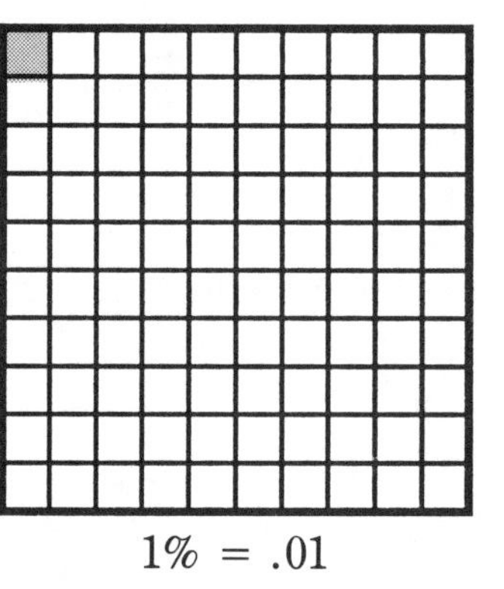

1% = .01

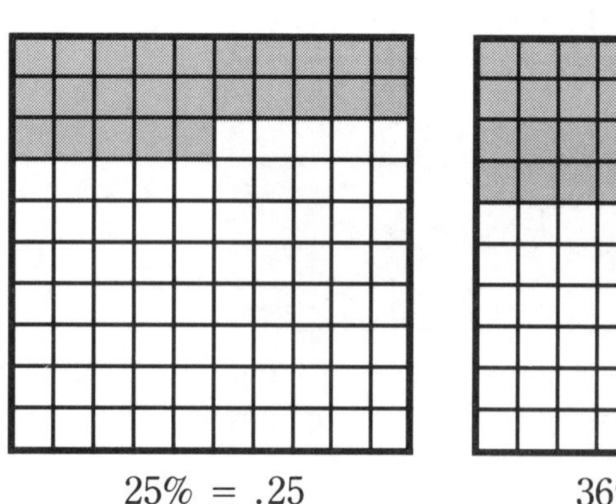

25% = .25

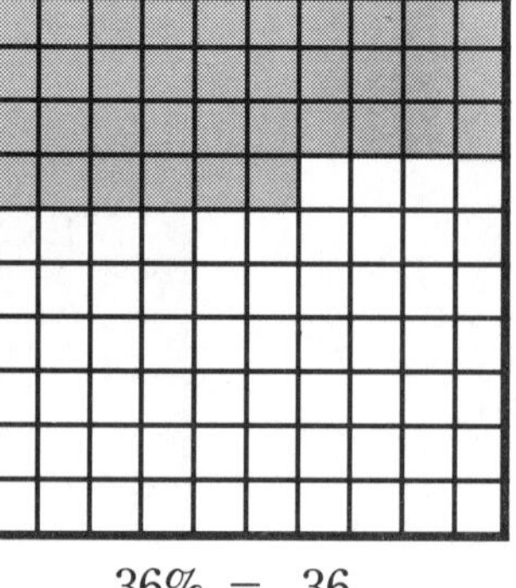

36% = .36

To change percents to decimals, follow the steps below. Then study the examples.

Change 15% to a decimal.

Step 1: Take away the percent sign. Place a decimal point to the right of the last digit. 15% 15.

Step 2: Move the decimal point two places to the left. .15

When the percent is a single digit, write a zero before moving the decimal point.

$$8\% = .08$$

When the percent is a fraction, write two zeros before placing the decimal point.

$$\frac{3}{4}\% = .00\frac{3}{4}$$

When the percent is a mixed number, write any necessary zeros before moving the decimal.

$$4\frac{1}{2}\% = .04\frac{1}{2}$$

PRACTICE

Change the percents below to decimals. Use a separate sheet of paper.

1. 32%
2. 21%
3. 43%
4. $\frac{1}{2}\%$
5. 7%
6. $5\frac{3}{4}\%$
7. 125%
8. 19%
9. 7.5%
10. $\frac{3}{5}\%$
11. 12.5%
12. $\frac{4}{5}\%$
13. 91%
14. 115%
15. 101%
16. 18.3%
17. 48%
18. 1%
19. 14.5%
20. $\frac{1}{5}\%$
21. 46.5%
22. 7%
23. 36%
24. $2\frac{1}{2}\%$
25. 250%
26. 80%
27. 100%
28. $5\frac{3}{8}\%$
29. 50%
30. 140.7%

11.4 Changing Decimals to Percents

Here is the method for changing decimals to percents.

Follow the steps below.

Change .37 into a percent.

Step 1: Move the decimal point *two* places to the right. .37 = 37.

Step 2: Write a percent sign in place of the decimal point. 37%

If the decimal is a whole number, write two zeros before moving the decimal. Write a percent sign.

6 = 6.00 = 600%

Remember to write in any needed zeros to fill out places when moving the decimal.

14.5 = 14.50 = 1450%

PRACTICE

Change the decimals below to percents. Use a separate sheet of paper.

1. .5
2. .16
3. .32
4. .125
5. 17
6. .80
7. 4.38
8. .527
9. .89
10. 12.4
11. .3
12. .103
13. 5.28
14. .25
15. 5.7
16. 2.666

11.5 Changing Percents to Fractions

There are several methods for changing percents to fractions. Follow the steps below.

Use this method when the percent is a whole number: 6%

Step 1: Write the percent as a fraction with the denominator 1. Drop the % sign. $6\% = \frac{6}{1}$

Step 2: Multiply the fraction by 1/100. $\frac{6}{1} \times \frac{1}{100} = \frac{6}{100}$

Step 3: Reduce the answer to lowest terms if possible. $\frac{6}{100} = \frac{3}{50}$

The above steps also apply if the whole number is larger than 100. For example: Change 140% to a fraction.

$$140\% = 140$$

$$\frac{140}{1} \times \frac{1}{100} = \frac{140}{100}$$

$$\frac{140}{100} = 1\frac{40}{100} = 1\frac{4}{10} = 1\frac{2}{5}$$

Use this method when the percent is a fraction: $\frac{3}{4}\%$

Step 1: Drop the % sign.

Step 2: Multiply by 1/100. $\frac{3}{4} \times \frac{1}{100} = \frac{3}{400}$

Step 3: Reduce the answer to lowest terms if possible.

Use this method when the percent is a mixed number: $7\frac{1}{2}\%$

Step 1: Drop the % sign. Change the mixed number to an improper fraction. $7\frac{1}{2} = \frac{15}{2}$

Step 2: Multiply by 1/100. $\frac{15}{2} \times \frac{1}{100} = \frac{15}{200}$

Step 3: Reduce the answer to lowest terms if possible. $\frac{15}{2} \times \frac{1}{100} = \frac{15}{200} = \frac{3}{40}$

PRACTICE

Change the percents below to fractions. Do your work on a separate sheet of paper.

1. 4% **5.** 335% **9.** 150% **13.** 50%

2. 25% **6.** 400% **10.** $12\frac{1}{3}\%$ **14.** $30\frac{1}{4}\%$

3. 80% **7.** $6\frac{3}{4}\%$ **11.** 375% **15.** 80%

4. 33% **8.** $9\frac{1}{2}\%$ **12.** 15% **16.** 200%

11.6 Changing Fractions and Mixed Numbers to Percents

Here is the method for changing fractions to percents. Follow the steps below.

For example: Write $\frac{2}{3}$ as a percent.

Step 1: Divide the numerator by the denominator. Add a decimal point and two zeros to the dividend.

$$\frac{2}{3} = 3\overline{)2.00}$$

Step 2: Find the quotient to two decimal places. Then remove the decimal point and write a percent sign after the number.

$$\begin{array}{r} .66\frac{2}{3} \\ 3\overline{)2.00} \\ \underline{1\,8} \\ 20 \\ \underline{18} \\ 2 \end{array}$$

$$.66\tfrac{2}{3} = 66\tfrac{2}{3}\%$$

Changing mixed numbers to percents is almost as easy as changing fractions to percents. Follow the steps below.

For example: Write $4\frac{3}{4}$ as a percent.

Step 1: Change the mixed number to an improper fraction.

$$4\frac{3}{4} = \frac{19}{4}$$

Step 2: Divide the numerator by the denominator. Add a decimal point and two zeros to the dividend.

$$4\overline{)19.00}$$

Step 3: Find the quotient to two decimal places. Then remove the decimal point and write a percent sign after the number.

$$\begin{array}{r} 4.75 \\ 4\overline{)19.00} \\ \underline{16} \\ 30 \\ \underline{28} \\ 20 \\ \underline{\underline{20}} \end{array}$$

$$4.75 = 475\%$$

PRACTICE

A. Change the fractions and mixed numbers below to percents. Use a separate sheet of paper.

1. $\frac{1}{4}$	**3.** $\frac{1}{200}$	**5.** $\frac{7}{35}$	**7.** $5\frac{3}{8}$	**9.** $7\frac{1}{10}$
2. $\frac{3}{5}$	**4.** $\frac{6}{9}$	**6.** $1\frac{2}{3}$	**8.** $\frac{27}{300}$	**10.** $14\frac{5}{6}$

B. Copy the chart below on a separate sheet of paper. Each row has one number in either the percent, decimal, or fraction column. Fill in the other two columns in each row. Work on a separate sheet of paper.

Percent	Decimal	Fraction
1) 5%		
2)	.25	
3)		$\frac{3}{4}$
4)	.40	
5) 32%		

11.7 Comparing Percents, Decimals, and Fractions

You can compare different percents, decimals, and fractions. Follow the steps below. For example: Compare 40%, .80, and $\frac{3}{5}$.

Step 1: Change all the numbers to decimals or fractions.

Changing to Decimals

40% $\quad$ 40.% $\quad = .40$

.80 $\quad = .80$

$\frac{3}{5}$ $\quad$ $5\overline{)3.00}$ gives $.60$; $3\,0$; 0 $\quad = .60$

Changing to Fractions

40% $\quad \frac{40}{1} \times \frac{1}{100} = \frac{40}{100} = \frac{2}{5}$

.80 $\quad \frac{80}{100} = \frac{4}{5}$

$\frac{3}{5} = \frac{3}{5}$

Notice that the method for changing fractions to decimals is nearly the same as changing fractions to percents. Divide the numerator by the denominator. Add a decimal point and two zeros to the dividend. Then write the quotient as a decimal.

Step 2: Compare place value. Check to see that the fractions have a common denominator. If they do not, change the fractions so that they do. Compare numerators.

Step 3: Write the numbers in order of increasing value.

40% $\frac{3}{5}$ .80

PRACTICE

Compare the numbers below. Write them in order of increasing value.

Remember: Before you can compare these numbers, change them all to decimals or fractions. Use a separate sheet of paper.

1. 15% $\frac{1}{10}$ .16

2. 100% $\frac{9}{10}$ .75

3. 22% $\frac{3}{5}$ .05

4. 35% $\frac{3}{4}$ .66

11.8 Finding the Percent of a Number

The method for finding the percent of a number is simple.

For example: Find 32% of 78 =

Follow the steps below.

Step 1: Change the percent to a decimal.

32.% = .32

Step 2: Multiply the other number in the problem by the decimal.

Step 3: Put the decimal point in the correct place in the product.

$$\begin{array}{r} 78 \\ \times .32 \\ \hline 1\,56 \\ 23\,4 \\ \hline 24.96 \end{array}$$

32% of 78 = 24.96

PRACTICE

Solve the problems below on a separate sheet of paper.

1. 10% of 60 =

2. 18% of 40 =

3. 25% of 80 =

4. 12% of 58 =

5. 15% of 95 =

6. 9% of 81 =

7. 56% of 84 =

8. 75% of 96 =

9. 34% of 100 =

10. 27% of 800 =

11. 62% of 224 =

12. 49% of 1,000 =

11.9 Finding a Number When a Percent of It Is Known

Here is the method for finding a number when a percent of it is known.

For example: 20% of (*what number*) is 5?

Follow the steps below.

Step 1: Change the percent to a decimal.

20.% = .20

Step 2: Divide the other number in the problem by the decimal.

20% of *25* = 5

$$\begin{array}{r} 25 \\ .20\overline{)5.00} \\ \underline{4\,0} \\ 1\,00 \\ \underline{1\,00} \end{array}$$

PRACTICE

Solve the problems below on a separate sheet of paper.

1. 10% of ____ = 8

2. 15% of ____ = 40

3. 5% of ____ = 50

4. 9% of ____ = 20

5. 12% of ____ = 64

6. 20% of ____ = 32

7. 74% of ____ = 102

8. 66% of ____ = 34

9. 23% of ____ = 100

10. 60% of ____ = 80

11. 50% of ____ = 250

12. 49% of ____ = 94

11.10 Finding What Percent One Number Is of Another

Sometimes you may need to find what percent one number is of another.

For example: What percent of 50 is 10?

Follow the steps below.

Step 1: Express both numbers as a fraction. Make the number that represents the whole part the denominator. Make the number that represents the percent the numerator.

$$\frac{10}{50}$$

Step 2: Divide the numerator by the denominator. Add a decimal point and two zeros to the dividend.

$$50\overline{)10.00} = .20$$
$$\underline{10\ 0}$$

Step 3: Write the decimal quotient as a percent.

.20 = 20%

> If one of the numbers is a decimal, do not express the numbers as a fraction. Divide the number that represents the percent by the number that represents the whole part. Make sure you place the decimal point in the quotient directly over the one in the dividend.
>
> For example: What percent of 55 is 6.6?
>
> $$55\overline{)6.60} = .12$$
> $$\underline{5\ 5}$$
> $$1\ 10$$
> $$\underline{1\ 10}$$
>
> .12 = 12%

PRACTICE

Solve the problems below on a separate sheet of paper.

1. What percent of 24 is 12?
2. What percent of 64 is 8?
3. What percent of 96 is 4?
4. What percent of 200 is 20?
5. What percent of 35 is 7?
6. What percent of 57 is 5?
7. What percent of 125 is 9?
8. What percent of 38 is 2.4?
9. What percent of 69 is 3.5?
10. What percent of 42 is .8?

11.11 Estimating Percents

Sometimes you may need to estimate the answer to a problem about percents.

When you do estimate, remember two things.

1. Notice the percent of the number you are trying to find. If it is less than 100%, the answer will be *smaller* than the whole you are working with. For example: Find 80% of 200.

   percent   whole

   80% of 200 is *160.*

2. Your answer should make sense. You know that 95% is closer to 100% than 30% is. For example: Find 95% of 60; find 30% of 60.

   percent   whole

   95% of 60 is 57.

   percent   whole

   30% of 60 is 18.

95% of 60 is closer to 60 than 30% of 60 is.

57 is closer to 60 than 18 is.

PRACTICE

Read each problem below. Choose the exact answer or the one that is closest to the exact answer. Write the letter of that answer on a separate sheet of paper.

1. 25% of 50 is	**a.** 90	**b.** 40	**c.** 20
2. 10% of 40 is	**a.** 32	**b.** 4	**c.** 40
3. 55% of 250 is	**a.** 140	**b.** 200	**c.** 3.50
4. 12% of 36 is	**a.** 12	**b.** 32	**c.** 4.32

11.12 Solving Word Problems with Percent

Steps in Solving Word Problems

Read this problem. Follow the steps below to solve it.

Logan's Furniture Store was having a sale. Everything was 30% off the list price. Jed and Grace found a sofa they liked. Its list price was $795. What would the sale price of the sofa be?

Step 1: Read the problem.

Step 2: Learn what you must find out. 1) How much is 30% of $795? 2) What will the price be after this amount is subtracted from $795?

Step 3: Notice the numbers. Which number is the whole? What is the percent?

The whole: $795
The percent: 30%

Step 4: Find how much 30% of $795 is.

$$\begin{array}{r} \$795 \\ \times\ .30 \\ \hline \$238.50 \end{array}$$

Step 5: Subtract that amount from $795.

$$\begin{array}{r} \$795.00 \\ -\ 238.50 \\ \hline \$556.50 \end{array}$$

Step 6: Check the answer. Ask yourself if it makes sense.

WORD PROBLEMS TO SOLVE

Solve the problems below on a separate sheet of paper.

1. Don got 20% off on the stereo he bought. The original cost of the stereo was $462.50. How much did Don pay for the stereo?
2. Sandra and Tony buy unpainted furniture. Then they finish it the way a customer wishes and resell it. They bought a chest of drawers for $75.95. They resold it for 175% of that price. For how much did they sell it? How much money did they make?
3. A factory has 258 workers. A total of 42% of them drive to work. How many workers at this factory drive to work?
4. There are 850 workers at another factory who drive to work. A total of 25% of them are in car pools. How many workers are in car pools?
5. A clothing store had a sale. Every item cost $33\frac{1}{3}\%$ less than the price on the ticket. John bought a sweater with a ticket price of $45.00. How much did he pay for the sweater?
6. Harry lost 40% of his tools in a garage fire. He used to have 37 tools. How many tools did Harry lose?
7. Tammy answered 96% of the problems on her English test? The test had 200 problems. How many problems did Tammy answer?
8. Derrick has saved 80% of the money he needs to buy a radio. The radio costs $42. How much money has Derrick saved?
9. There are 900 students in the senior class at Kennedy High School. 62% of them are going to the senior prom. How many seniors are going to the senior prom?
10. Matt's take home salary is $1200 per month. 35% of his salary goes to pay his rent. How much does Matt pay for his monthly rent?

MATHEMATICS IN YOUR LIFE: Saving Money

Doris worked at the Central City bank. It was her job to help customers open accounts.

"There are different ways of saving money at this bank," she explained to Marge. "You could put your money in a passbook account. Or you could put your money in a savings certificate instead."

"What's the difference?" asked Marge.

"A passbook account pays less interest. But you can take out money at any time," answered Doris. "When you put money in a savings certificate there are certain rules. You agree to leave your money in the bank for a certain amount of time. You have to pay a penalty if you take your money out earlier. But the certificate pays more interest. This chart will show you."

Length of Time	Interest Rate
3-month certificate	5.8%
6-month certificate	6.2%
9-month certificate	6.6%
12-month certificate	7%

Marge has $1,200.

1. How much interest would she get on her money if she put it into a 3-month certificate?
2. How much interest would she get on her money if she put it into a 6-month certificate?
3. How much interest would she get on her money if she put it into a 9-month certificate?

4. How much interest would she get altogether if she put $500 into a 6-month certificate and $700 into a 12-month certificate?
5. Look at your four answers. Which shows the most interest?

11.13 Using Your Calculator: Percents

You can use your calculator to work with percents. Follow the steps below.

Find 17% of 524

Step 1: Change the percent to a decimal. Take off the percent sign. Move the decimal point two places left.

17% = .17

Step 2: Press [5] [2] [4]

Step 3: Press [×]

Step 4: Press [·] [1] [7]

If your calculator has a percent sign, don't change the percent to a decimal. In *Step 4* press 17 and the percent key.

Step 5: Press [=]

Step 6: Your answer should be: 89.08

PRACTICE

Use your calculator to solve these problems.

1. 32% of ____ = 84
2. 74% of 496 =
3. 29% of ____ = 362
4. 82% of 685 =
5. 31% of ____ = 493
6. 24% of 762 =

CHAPTER REVIEW

CHAPTER SUMMARY

■ Percent %	A percent is a part of a whole that has been divided into 100 equal parts. The sign for percent is %. A whole is 100%. Any percent number less than 100% is less than a whole. Any percent number more than 100% is more than a whole.	7% 120% $3\frac{1}{2}$%
■ Changing Percents	Percents may be changed to decimals and fractions.	25% = .25 = $\frac{1}{4}$
■ Comparing Percents	The value of percents, decimals, and fractions can be compared.	From smallest to largest: 25% .50 $\frac{3}{4}$
■ Working with Percents	Any percent of a number can be found.	
	A number can be found when a percent of it is known.	25% of *40* = 10
	The percent that one number is of another can be found.	25% of 20 = *5*

REVIEWING VOCABULARY

Use the correct words from the box to complete the sentences. Write on a separate sheet of paper.

percent	compared	decimal	estimate	fraction

1. A ________ number is written with a decimal point followed by places to the right.

2. A ________ has a numerator and a denominator.

3. The ________ sign is written as %.

4. The value of percents, decimals and fractions can be ________.

5. The ________ using percents helps you to guess an answer.

Chapter Review

Chapter Quiz

A. Change each percent below to a fraction and a decimal. Use a separate sheet of paper.

1. 4%
2. 65%
3. $\frac{2}{5}\%$
4. $8\frac{3}{4}\%$
5. $52\frac{1}{2}\%$
6. 12%
7. 86%
8. $\frac{5}{8}\%$
9. $28\frac{2}{3}\%$

B. Solve the problems below. Use a separate sheet of paper.

1. What percent of 94 = 3?
2. What percent of 72 = 4?
3. 16% of ________ = 4.
4. 22% of ________ = 6.
5. 32% of 120 = ?
6. What percent of 80 = 5?
7. What percent of 57 = 3?
8. 35% of ________ = 28.
9. 45% of 90 = ?
10. 58% of 200 = ?

C. Solve these word problems on a separate sheet of paper.

1. Laurie wants to buy a car that costs $8,825. She must make a down payment of 28%. How much will her down payment be? How much money will she have to pay after the down payment?
2. Ron gets a 15% commission on everything he sells. Last month he sold $6,327 worth of goods. What was his commission?
3. Eileen bought a suitcase for $64.98. The sales tax was $6\frac{1}{2}\%$. How much was the tax? What was the total cost of the suitcase?
4. 18% of Joanna's salary goes to pay her utility bills. If she makes $1800 per month, how much does she spend on her utilities?
5. Mr. Bennett has 35 students in his fourth period math class. 20% of these students received a "B" in this class. How many students received a grade of "B"?

RATIOS AND PROPORTIONS

12

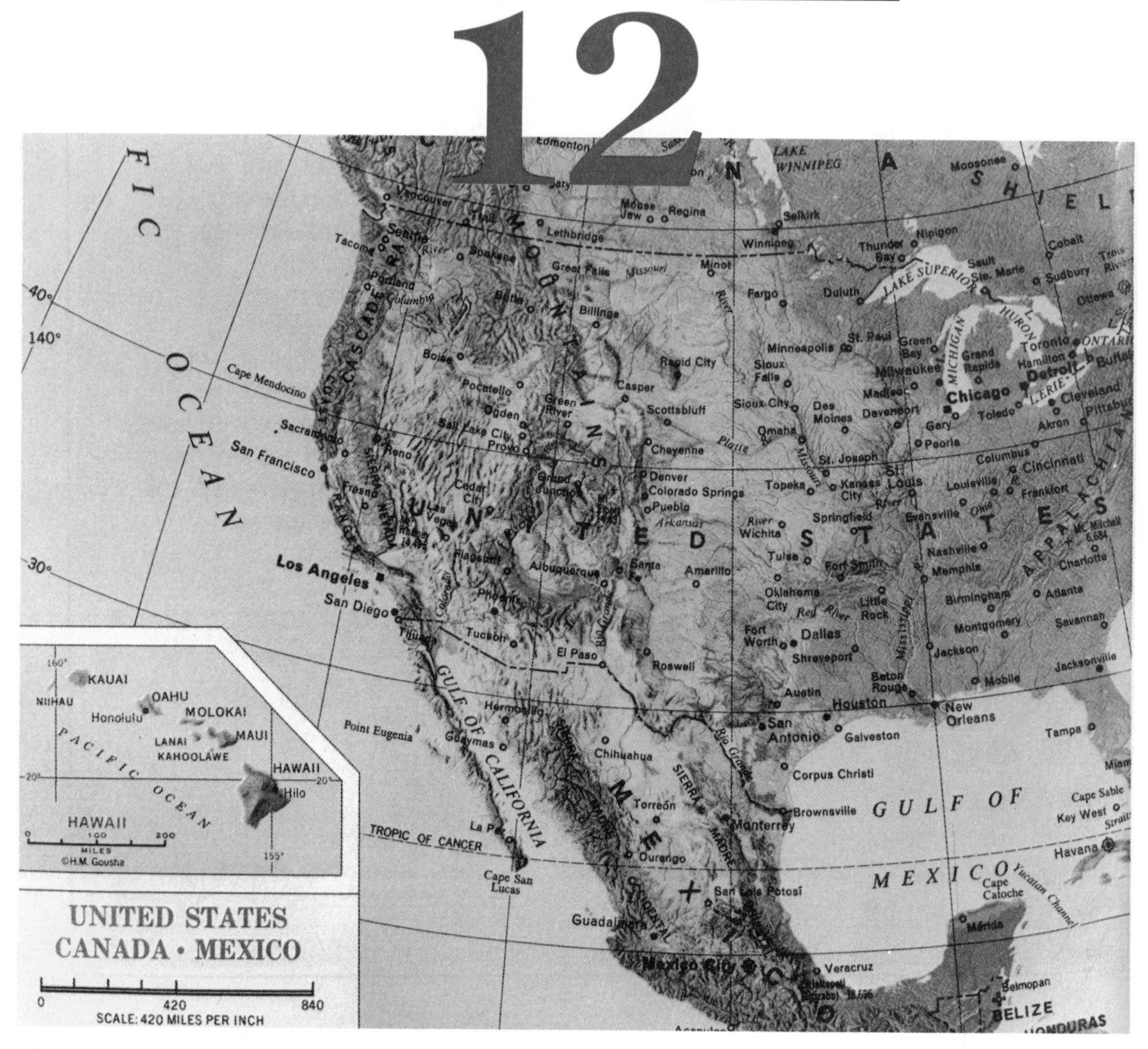

The map in this photograph has a scale of 420 miles per inch. That means the ratio of inches to miles is 1 to 420. This chapter will show many other ways to use ratios and proportions in numbers.

Chapter Learning Objectives

1. Understand the meaning of ratio
2. Understand the meaning of proportion
3. Identify proportions
4. Solve proportions
5. Solve problems with proportions

WORDS TO KNOW

■ ratio	a comparison of one number to another
■ proportion	a statement that two ratios are equal
■ term	any of the numbers in a ratio or proportion
■ ratio sign :	a colon used between two numbers to show a ratio
■ proportion sign ::	a double colon used between two ratios to show that they are equal
■ cross multiply	to multiply the numerator of one fraction by the denominator of the other
■ cross product	the result of cross multiplying

12.1 What Is a Ratio?

A ratio is a comparison of one number to another. Those numbers are called terms.

A ratio can be described in many ways.

In words:
3 buses to 120 students
3 buses for 120 students
3 buses per 120 students

With a ratio sign: 3 buses : 120 students
3 : 120

As fractions:

$$\frac{3 \text{ buses}}{120 \text{ students}}$$

$$\frac{1}{40}$$

All of the above express the ratio $\frac{3}{120}$.

Notice that the first term, 3, is the numerator of the fraction. The second term, 120, is the denominator.

The ratio in fraction form tells us how many *times* the first term is bigger (or smaller) than the second.

In the above example $\frac{3}{120} = \frac{1}{40}$.

3 is smaller than 120. 3 is $\frac{1}{40}$ of (times) 120. Check: $3 = \frac{1}{40} \times 120$.

PRACTICE

On a separate sheet of paper, write the ratio for each phrase below in three ways. First write the ratio in words. Next write the ratio as a fraction, and finally write it with the ratio sign.

1. 1 inch to 14 miles
2. 12 eggs to a carton
3. 62 miles in 2 hours
4. 93 people for 98 seats
5. 80 people in 3 buses
6. 3.5 yards for 1 dress
7. 53 chairs for 7 tables
8. 5 shelves for 138 books

12.2 What Is a Proportion?

A proportion is a statement that two ratios are equal. That means the fractions are equivalent.

To review equivalent fractions, turn to page 115.

Look at the left square below. One of the two parts is in color, a ratio of 1:2 or $\frac{1}{2}$.

Look at the right square below. Four of the eight parts are in color, a ratio of 4:8 or $\frac{4}{8}$.

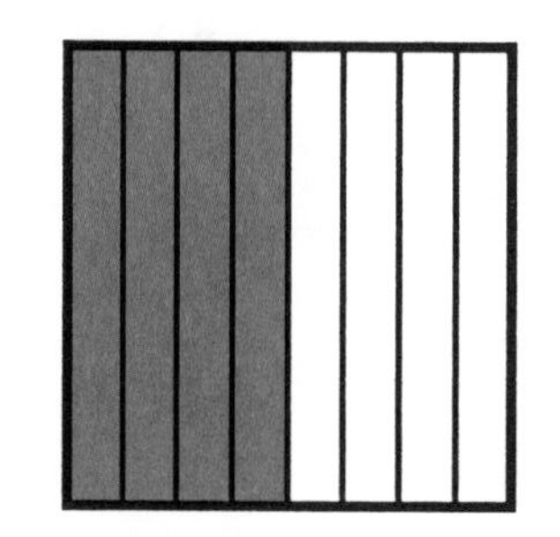

$\frac{1}{2} = \frac{4}{8}$, so the ratio of colored to uncolored parts is the same for each square.

This proportion, $\frac{1}{2} = \frac{4}{8}$, can also be written:

One is to two as four is to eight or 1:2 :: 4:8. The proportion sign, :: , means that the two ratios are equal.

PRACTICE

A. Express the proportions below as fractions. Write your answers on a separate sheet of paper.

1. 10 : 20 :: 14 : 28
2. 12 : 36 :: 15 : 45
3. 18 : 54 :: 2 : 6
4. 7 : 63 :: 9 : 81
5. 36 : 48 :: 30 : 40
6. 10 : 100 :: 100 : 1,000
7. 35 : 105 :: 400 : 1,200
8. 60 : 90 :: 90 : 135
9. 25 : 125 :: 50 : 250
10. 90 : 100 :: 45 : 50

B. Each set of squares below shows a proportion.

The proportion shown in the squares is the same as the unfinished proportion written below it. Finish writing the proportions on a separate sheet of paper.

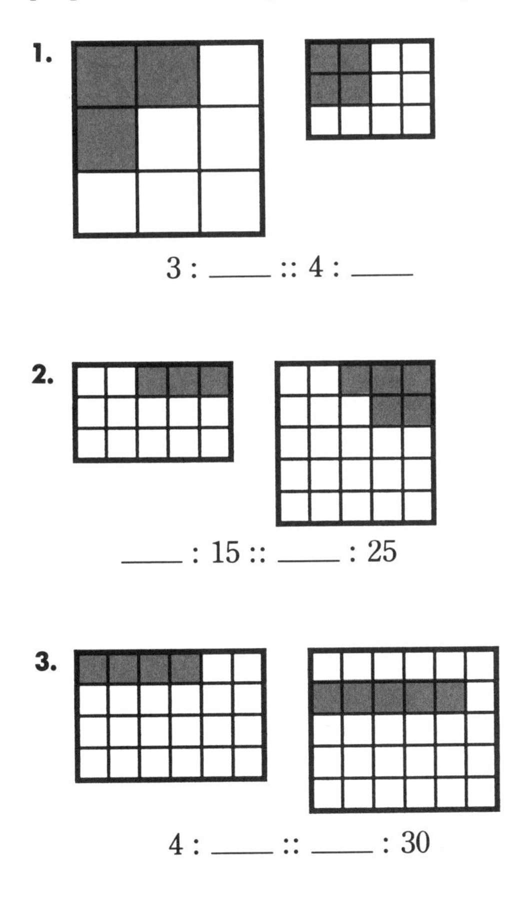

4. 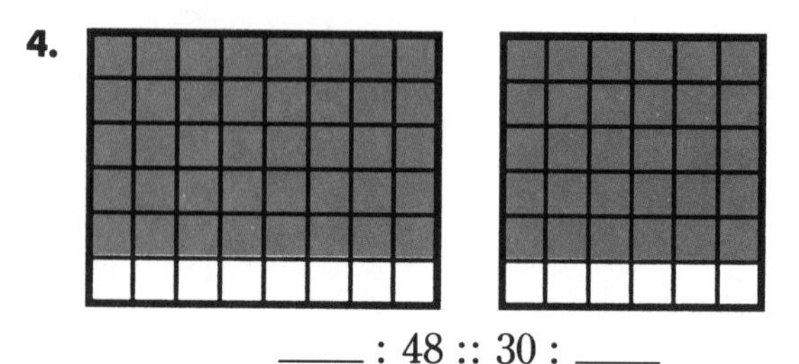

____ : 48 :: 30 : ____

C. On a separate sheet of paper, write the proportions below as fractions. Then write them using the proportion sign.

1. 6 is to 9 as 12 is to 18
2. 3 is to 12 as 4 is to 16
3. 9 is to 27 as 16 is to 48
4. 55 is to 110 as 22 is to 44
5. 24 is to 36 as 70 is to 105

12.3 Equal Ratios and Proportions

You have already learned that when two ratios are equal they form a proportion. There is a simple way of finding out if two ratios are equal. Follow the steps below.

Do $\frac{3}{4}$ and $\frac{12}{16}$ form a proportion?

Are $\frac{3}{4}$ and $\frac{12}{16}$ equal?

Step 1: Cross multiply the denominator of $\frac{3}{4}$ by the numerator of $\frac{12}{16}$. $4 \times 12 = 48$

Step 2: Cross multiply the denominator of $\frac{12}{16}$ by the numerator of $\frac{3}{4}$. $16 \times 3 = 48$

Step 3: Compare the cross products. If they are equal, the ratios form a proportion.

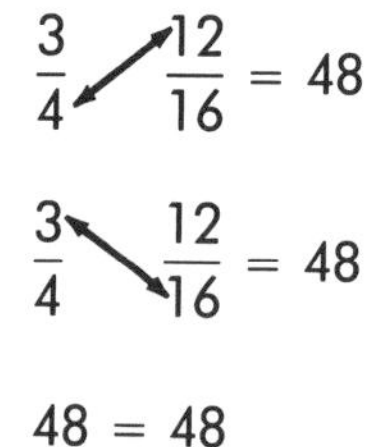

48 = 48
3 : 4 :: 12 : 16

Some ratios contain mixed numbers. You can also find out if these ratios are equal.

Do $\frac{2\frac{2}{3}}{4}$ and $\frac{4}{6}$ form a proportion?

Are $\frac{2\frac{2}{3}}{4}$ and $\frac{4}{6}$ equal?

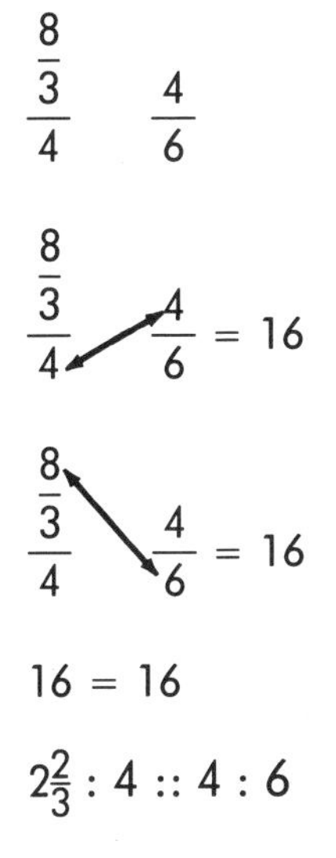

Step 1: Change $2\frac{2}{3}$ to an improper fraction.

$2\frac{2}{3} = \frac{8}{3}$

Step 2: Cross multiply the denominator 4 by the numerator 4. $4 \times 4 = 16$

Step 3: Cross multiply the denominator 6 by the numerator $\frac{8}{3}$. $\frac{8}{3} \times \frac{6}{1} = \frac{48}{3} = 16$

Step 4: Compare the cross products. If they are equal, the ratios form a proportion.

$16 = 16$

$2\frac{2}{3} : 4 :: 4 : 6$

PRACTICE

Find out if each pair of ratios below form a proportion. Cross multiply on a separate sheet of paper. Write = (equal) or ≠ (not equal).

1. $\frac{3}{4}$ $\frac{18}{24}$

2. $\frac{2}{7}$ $\frac{9}{10}$

3. $\frac{3}{9}$ $\frac{6\frac{1}{3}}{19}$

4. $\frac{5}{9}$ $\frac{4}{7}$

5. $\frac{\frac{2}{3}}{4}$ $\frac{6}{8}$

6. $\frac{\frac{1}{2}}{2}$ $\frac{4}{16}$

7. $\frac{5}{8}$ $\frac{2\frac{1}{2}}{4}$

8. $\frac{1\frac{2}{3}}{3}$ $\frac{10}{12}$

9. $\frac{2\frac{1}{4}}{9}$ $\frac{6\frac{1}{2}}{26}$

12.4 Solving Proportions

Sometimes you may have to find an unknown term in a proportion. Here is a method you can use to find this missing number. Follow the steps below.

Notice the number you divide by in Step 4.

Step 1: Let ? stand for the missing number.

$$\frac{2}{3} = \frac{?}{21}$$

Step 2: Cross multiply the numerator and denominator that are known.

$$2 \times 21 = 42$$

Step 3: Divide the product by the remaining number.

$$42 \div 3 = 14$$

Step 4: Write the dividend as the value of the missing number.

$$? = 14$$

Step 5: Check the ratios to see if they form a proportion.

$$\frac{2}{3} \times \frac{14}{21} \quad 42 = 42$$

$$\frac{2}{3} = \frac{14}{21}$$

PRACTICE

Find the missing number in each proportion below. Work on a separate sheet of paper. The first one has been done for you.

1. $\frac{?}{12} = \frac{5}{10}$

$5 \times 12 = 60$

$60 \div 10 = 6$

$? = 6$

$\frac{6}{12} = \frac{5}{10}$

2. $\frac{4}{?} = \frac{7}{21}$

3. $\frac{3}{4} = \frac{?}{16}$

4. $\frac{7}{9} = \frac{4}{?}$

5. $\frac{2}{9} = \frac{?}{21}$

6. $\frac{?}{18} = \frac{4}{24}$

7. $\frac{8}{32} = \frac{?}{16}$

8. $\frac{3}{?} = \frac{5}{15}$

9. $\frac{\frac{8}{12}}{12} = \frac{?}{14}$

10. $\frac{1\frac{1}{3}}{3} = \frac{?}{8}$

11. $\frac{\frac{3}{5}}{8} = \frac{2}{?}$

12. $\frac{\frac{1}{3}}{6} = \frac{?}{9}$

12.5 Solving Problems with Proportions

Read this problem. Use the method below to solve it.

Scottsville has 1 public library for every 6,000 people. The town wants to keep that proportion. How many libraries will Scottsville need when it has 24,000 people?

Steps in Solving Word Problems

Step 1: Read the problem.

$$\frac{1}{6{,}000} = \frac{?}{24{,}000}$$

Step 2: Write the proportion. Let ? stand for the missing number.

Step 3: Cross multiply the numerator and denominator that are known.

$$24{,}000 \times 1 = 24{,}000$$

Step 4: Divide the product by the remaining number.

$$24{,}000 \div 6{,}000 = 4$$

Step 5: Write the dividend as the value of the missing number.

$$? = 4$$

Step 6: Check the ratios to see if they form a proportion.

$$\frac{1}{6{,}000} \times \frac{4}{24{,}000}$$

$$24{,}000 = 24{,}000$$

$$\frac{1}{6{,}000} = \frac{4}{24{,}000}$$

PROBLEMS TO SOLVE

Use proportions to solve these word problems. Do your work on a separate sheet of paper.

1. Pink paint can be made by mixing red paint with white paint. Jim wants to get a certain shade of pink paint. To do this he needs to mix 1 quart of red paint with 3 quarts of white paint. To get the same color, how many quarts of red paint should he mix with 9 quarts of white?
2. Norma drove 660 miles. She used 33 gallons of gas. How many gallons will she need to drive 840 miles?

3. At Northside High School there are 2 staff members for every 36 students. Northside has 484 students. How many staff members are at Northside?
4. A map is drawn to a scale of 1 inch for every 15 miles. South City and Jacksonville are 12 inches apart. How many miles apart are they?
5. A train traveled 95 miles in 2 hours. How far can it travel in 5 hours?

MATHEMATICS IN YOUR LIFE: Prices and Proportions

Top Goods market was selling three heads of lettuce for \$1.55.

Janice needed two heads of lettuce for the salad she was going to make. She used proportion to find the cost of two heads of lettuce.

$$\frac{3}{1.55} = \frac{2}{?}$$

$2 \times 1.55 = 3.10$
$3.10 \div 3 = 1.033$
$? = 1.033$

Janice rounded \$1.033 to get the true cost of \$1.03.

PRACTICE

Solve the problems below on a separate sheet of paper.

1. Three cans of beans cost \$1.20. How much do five cans cost?
2. Two bunches of broccoli cost \$1.52. How much do three bunches cost?
3. Six cans of soup cost \$2.18. How much do four cans cost?

12.6 Using Your Calculator: Solving Proportions

You can solve proportions easily on your calculator.

Follow the steps to solve this problem:

The Hilldale College baseball team won 3 of its first 4 games. The team is scheduled to play 28 games for the season. Suppose the team continues to win games at the same rate. How many games will it have won by the end of the season?

Step 1: Press [3] [×] [2] [8] $\frac{3}{4} = \frac{?}{28}$

Step 2: Press [÷] [4] $3 \times 28 = 84$
$84 \div 4 = 21$

Step 3: Your answer should be: 21 $? = 21$

Step 4: Check the ratios to see if they form a proportion. $\frac{3}{4} = \frac{21}{28}$

PRACTICE

Use your calculator to solve the problems below. Write your answers on a separate sheet of paper. If your answer has more than two decimal places, round to the nearest hundredth.

1. $\frac{55}{120} = \frac{?}{205}$

2. $\frac{32}{?} = \frac{88}{112}$

3. $\frac{99}{100} = \frac{50}{?}$

4. $\frac{160}{200} = \frac{?}{340}$

5. $\frac{?}{500} = \frac{9}{10}$

6. $\frac{42}{?} = \frac{88}{420}$

7. $\frac{36}{250} = \frac{?}{1{,}000}$

8. $\frac{50}{?} = \frac{400}{640}$

9. $\frac{?}{155} = \frac{55}{775}$

10. $\frac{62}{125} = \frac{496}{?}$

11. $\frac{?}{375} = \frac{96}{2{,}250}$

12. $\frac{24}{360} = \frac{?}{45}$

CHAPTER REVIEW

CHAPTER SUMMARY

■ **Ratio**	A ratio is a comparison of one number to another. It can be expressed in words, as a fraction, or in numbers with the ratio sign.	3 to 4 $\frac{3}{4}$ $3 : 4$
■ **Proportion**	A proportion is a statement that two ratios are equal; the fractions are equivalent.	$3 : 4 :: 6 : 8$ $\frac{3}{4} = \frac{6}{8}$
■ **Cross Multiply**	To cross multiply, multiply each of the numerators of a pair of fractions by the denominator of the other. When equal ratios are cross multiplied, the products are the same. The missing term of a ratio can be found by cross multiplying, and then dividing.	$\frac{3}{4} \times \frac{6}{8}$ $24 = 24$ $\frac{2}{6} = \frac{?}{15}$ $2 \times 15 = 30$ $30 \div 6 = 5$ $? = 5$ $\frac{2}{6} = \frac{5}{15}$

REVIEWING VOCABULARY

Use the correct word or group of words from the box to complete the sentences. Use a separate sheet of paper.

ratio	proportion	term	cross multiply	cross product

1. A ________ states that two ratios are equal.
2. You can ________ to find the missing number of a proportion.
3. A ________ is the comparison of one number to another.
4. When you cross multiply you get a ________ .
5. A number in a ratio or a proportion is called a ________ .

CHAPTER REVIEW

CHAPTER QUIZ

A. Cross multiply to find if each pair of ratios is equal. Do your work on a separate sheet of paper.

1. $\frac{\frac{2}{3}}{6}$ $\frac{2}{18}$

2. $\frac{9}{12}$ $\frac{14}{42}$

3. $\frac{3}{5}$ $\frac{6}{8}$

4. $\frac{2\frac{1}{2}}{3}$ $\frac{7}{9}$

5. $\frac{\frac{7}{8}}{1}$ $\frac{11}{12}$

6. $\frac{5}{8}$ $\frac{1\frac{1}{2}}{2}$

B. Solve the problems below. Do your work on a separate sheet of paper.

1. $\frac{?}{3} = \frac{4}{5}$

2. $\frac{2}{?} = \frac{8}{14}$

3. $\frac{9}{18} = \frac{10}{?}$

4. $\frac{15}{27} = \frac{?}{9}$

5. $\frac{4\frac{1}{4}}{2} = \frac{?}{16}$

6. $\frac{?}{9} = \frac{5}{30}$

C. Developing Thinking Skills

Read the following paragraph. Then answer the questions below.

James is going to make a trip by car. He's going to travel from Kerryton to Midville. He wants to make the trip in one day. His map shows the distance between the two towns to be 12 inches. The scale of the map is one inch to 40 miles. James plans to leave his house at 9 a.m. He plans to make 2 stops for lunch and gas totalling $1\frac{1}{2}$ hours. He wants to be in Midville no later than 8 p.m. He knows the speed limit on the highway between the two towns is 55 mph. James' car holds 16 gallons of gas and he gets 20 miles per gallon.

1. How many miles is it from Kerryton to Midville?
2. Can James reach Midville by the time he wants without going over the speed limit?
3. If James follows his schedule, how many miles per hour will he have to average in order to reach Midville by 8 p.m.?
4. James will start his trip with a full tank of gas. How many gallons of gas will it take him to complete the trip?

UNIT THREE REVIEW

Copy this chart on a separate sheet of paper. Each row has one number in either the percent, decimal, or fraction column. Fill in the other two columns in each row.

	Percent	Decimal	Fraction
1.	6%		
2.		.85	
3.			$\frac{2}{5}$

4. Write the ratio below using the ratio sign.
2 feet to 18 yards

5. Write the proportion below using the proportion sign.
7 is to 14 as 5 is to 10.

6. On a separate sheet of paper, figure out if the ratios below form a proportion. Write $=$ or $\neq$.

$\frac{2}{5}$ and $\frac{8}{20}$

7. Find the unknown term in the proportion below.

$\frac{?}{18} = \frac{6}{12}$

Solve the problems below on a separate sheet of paper.

8.
$$\begin{array}{r} 23.035 \\ 72.4 \\ +46.75 \\ \hline \end{array}$$

9.
$$\begin{array}{r} 504.692 \\ -289.063 \\ \hline \end{array}$$

10. $.67\overline{)741}$

11.
$$\begin{array}{r} .93 \\ \times\ 10 \\ \hline \end{array}$$

12. $53\overline{)237.053}$

13. $2.93\overline{)1804.682}$

14. 30% of 120 =

15.
$$\begin{array}{r} .487 \\ \times\ 100 \\ \hline \end{array}$$

16. 5% of ____ = 27

UNIT FOUR

13
GRAPHS

14
STATISTICS AND PROBABILITY

GRAPHS

13

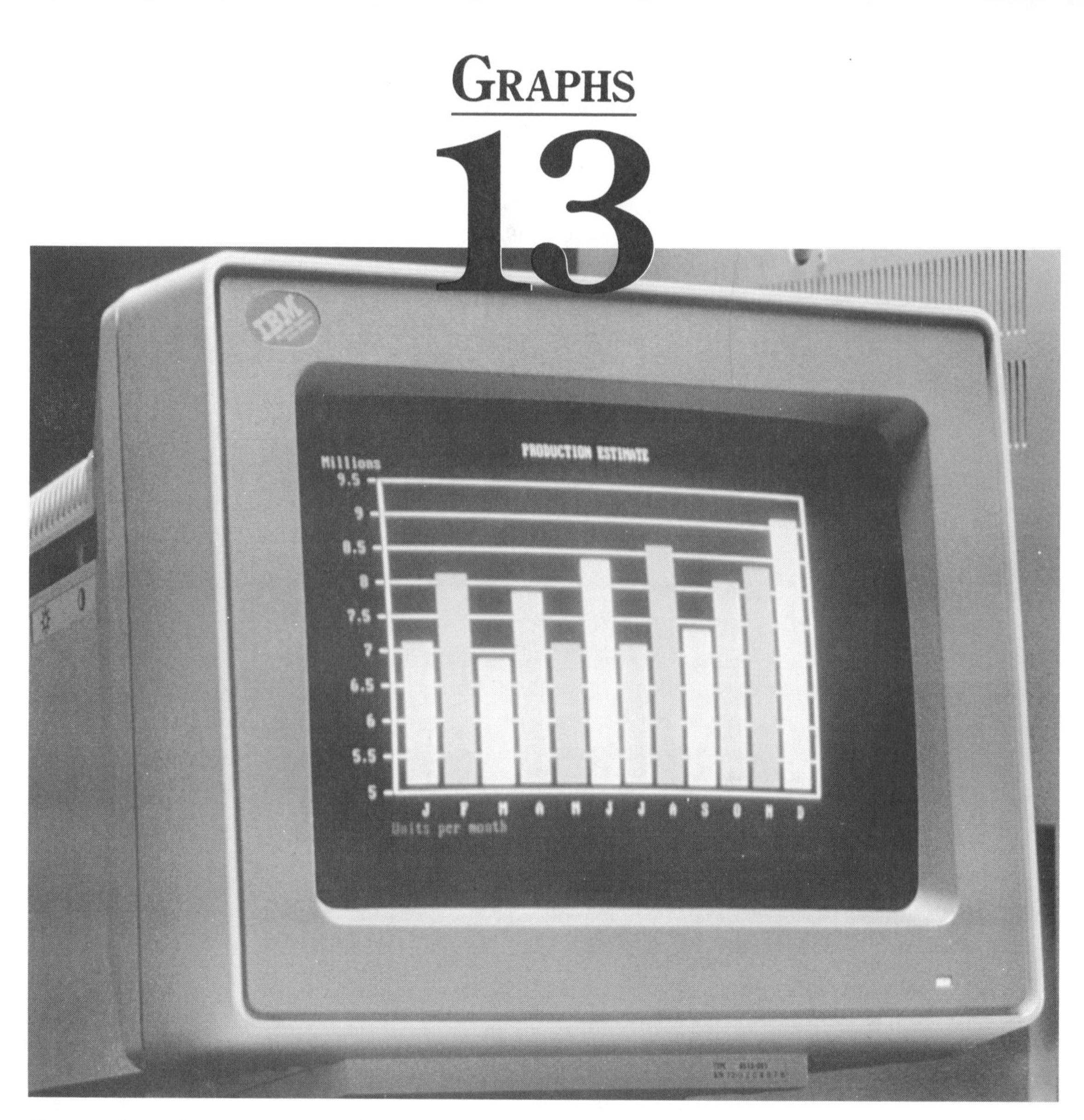

The graph on this computer is called a bar graph. A bar graph is one of three main graphs we use to help us understand and compare number facts. Can you remember a time when you read a graph to get some information you needed?

Chapter Learning Objectives

1. Read and construct bar graphs
2. Read and construct line graphs
3. Read and construct circle graphs

WORDS TO KNOW

- **graph** a picture that shows number facts
- **bar graph** a graph that uses bars of different lengths to compare amounts or sizes of things
- **line graph** a graph that uses line segments to show changes and relationships between things, usually over a period of time
- **circle graph** a circle-shaped graph that shows how a total amount (100%) has been divided into parts

A **graph** is a picture that shows information as number facts. There are three main types of graphs: bar graphs, line graphs, and circle graphs.

13.1 Bar Graphs

Bar graphs are used to compare amounts or sizes of things. Amounts on bar graphs are usually shown in round numbers.

This graph shows attendance at baseball games from 1983 to 1988.

Baseball Attendance

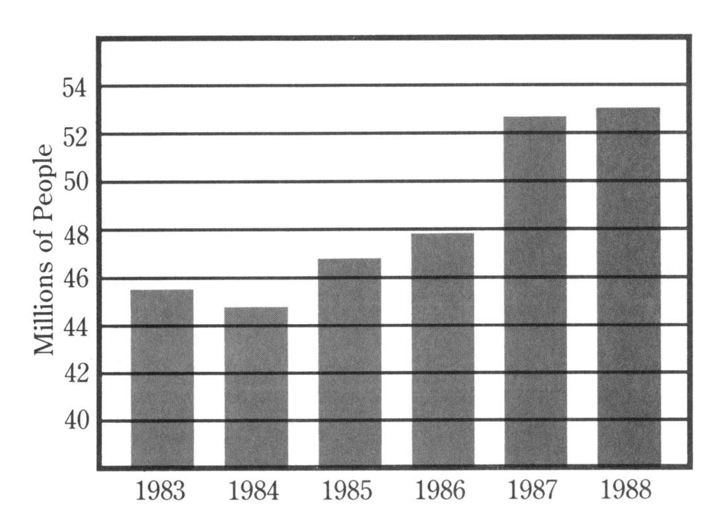

Look at the graph on the previous page. Notice the horizontal (straight across) lines. The space between each line represents a difference of two million fans. Now look at the vertical (up and down) bars. Each one represents fan attendance for a different year. Notice that most of the bars do not stop exactly on a horizontal line. Many bar graphs (and line graphs) show information that falls *in between* numbered lines. But you can estimate this information. You just need to figure out how close a bar (or point on a line graph) comes to a horizontal line.

For example, look at the baseball attendance for 1985. That bar reaches halfway between the horizontal lines for 46 and 48 million fans. That means approximately *47* million fans attended games that year. Now look at the bar for 1983. It reaches almost all the way to 46 million. That means you can estimate that approximately 45.7 or 45.8 million fans attended games that year.

All bar graphs and line graphs can be read this way. You can always estimate information between the horizontal lines.

More facts can be shown on a double or a triple graph.

This triple graph shows the percentage of games won by one high school's sports teams.

The graphs we have seen so far show baseball attendance and percentage of games won. What other kinds of sports information might you find on a bar graph?

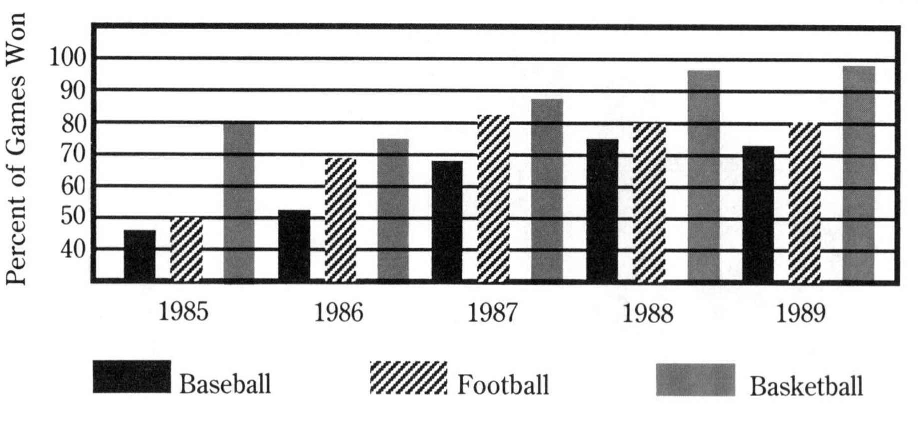

PRACTICE

A. Study the first bar graph. Answer these questions on a separate sheet of paper.

1. In what year was attendance at baseball games greatest? About how many millions of people attended baseball games that year?

2. In what year was the attendance smallest? About how many millions of people attended baseball games that year?

3. In which two years was there the greatest rise in attendance? About how many more people attended games in the second of the two years?

4. About how many fewer people attended games in 1984 than in 1983?

B. Study the second bar graph. Answer these questions on a separate sheet of paper.

1. Which 3 sports teams are shown on the graph? What color represents each team?

2. Which team had the best overall record?

3. Which team had the poorest overall record?

4. Which team made the greatest improvement from one year to the next? By what percent did they improve?

C. Make a bar graph that shows these facts. Work on a separate sheet of paper.

Attendance at the annual symphony concert had generally grown over the years.

In 1984 attendance was 3,500.

In 1985 attendance was 4,350.

In 1986 attendance was 4,100.

In 1987 attendance was 5,400.

In 1988 attendance was 6,200.

In 1989 attendance was 6,800.

13.2 Line Graphs

Line graphs are used to show changes and relationships between things, usually over a period of time.

This graph shows how much money was spent on construction from January through August in 1988. The amounts have been rounded to the nearest billion.

Construction Spending

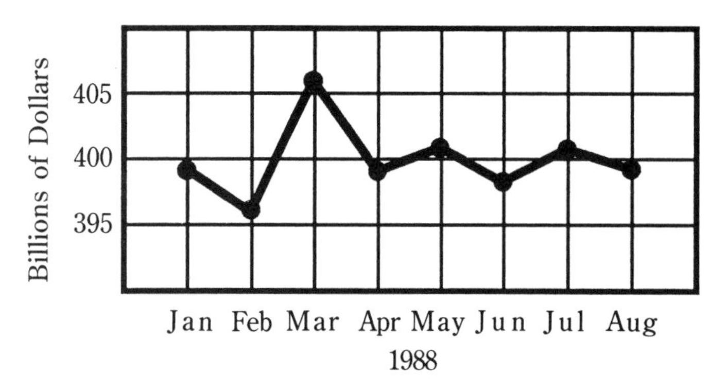

Line graphs can also be used to compare sets of related numbers. This graph shows the population changes in two cities.

Population of Hanley and Danton

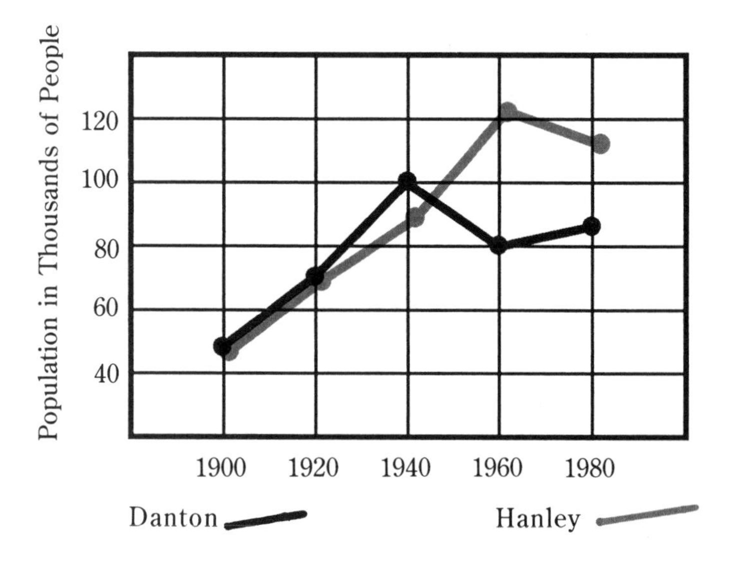

PRACTICE

A. Study the first graph. Answer these questions on a separate sheet of paper.

1. About how much money was spent in April?
2. In what month was the most money spent? About how much was spent in this month?
3. In what month was the least money spent? About how much was spent in this month?
4. About how much more money was spent in March than in February?

B. Study the second graph. Answer these questions on a separate sheet of paper.

1. Which line represents the population of Hanley? Which line represents the population of Danton?
2. Which city had the largest population in 1900? About how many more people did it have—2,000, 10,000, or 15,000?
3. In which year did Danton have its largest population?
4. About how many more people did Hanley have in 1980 than in 1900?
5. About how many more people did Danton have in 1980 than in 1900?

C. Show these facts in a line graph. Work on a separate sheet of paper.

The Apex Company sold 375 video cassettes in the month of January.

It sold 450 in February.

It sold 350 in March.

It sold 480 in April.

It sold 625 in May.

It sold 780 in June.

13.3 Circle Graphs

Circle graphs are used to show how the total amount (100%) of something is divided. You can compare amounts by comparing the sizes of the parts of the graph.

This circle graph shows where one city gets its money. The whole graph stands for one dollar. The parts of the graph show how many cents of each dollar come from each place. For example, 23¢ of every dollar comes from business tax. This means that 23% of the money comes from business tax.

Circle graphs are also called pie charts. Can you guess why?

Where the City Dollar Comes From

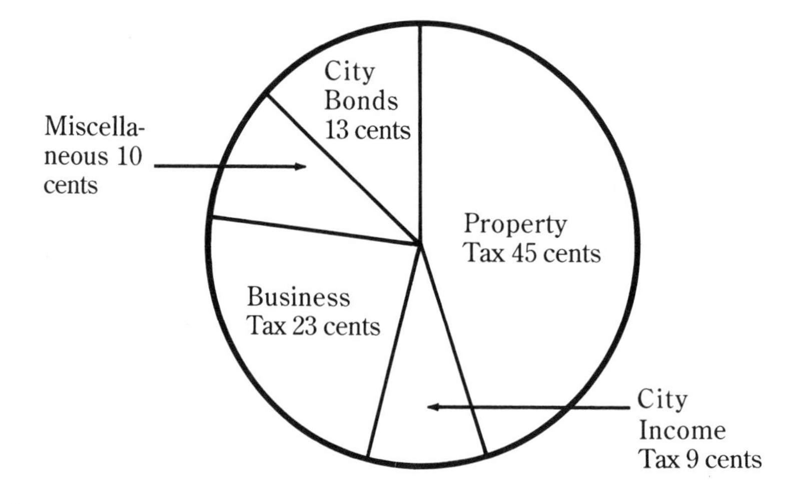

PRACTICE

A. Study the graph. Answer these questions on a separate sheet of paper.

1. From what source does the largest part of the city's revenue come? How much of each dollar is this?
2. From what source does the smallest part of the city's revenue come? What percent is this?
3. How much of every dollar comes from city bonds?
4. The city took in $900,000 last year. How much more money was raised from property tax than from business tax?

If you need help in finding the percent of a number, turn back to page 190.

B. Show these facts in a circle graph. Work on a separate sheet of paper.

In Center City people go to work in different ways.

45% of the people use buses.

28% of the people use the subway.

17% of the people drive alone.

9% of the people car pool.

1% of the people use other means.

C. Show these facts in a circle graph. Work on a separate sheet of paper.

A student's expenses must be carefully budgeted.

54% tuition

28% room and board

11% personal

5% books

2% travel

D. Show these facts in a circle graph. Work on a separate sheet of paper.

Bill gets money to buy a new car from several different sources.

45% from a car loan.

20% from the sale of his old car

25% from a gift from his grandfather

10% from his savings

E. Show these facts in a circle graph. Work on a separate sheet of paper.

Beth's Monthly Budget

36% rent

20% food

14% clothing

15% auto

15% miscellaneous

MATHEMATICS IN YOUR LIFE: Budgets

The Browns and the Petersons spend their incomes differently. Study the two circle graphs. Then answer the questions below on a separate sheet of paper.

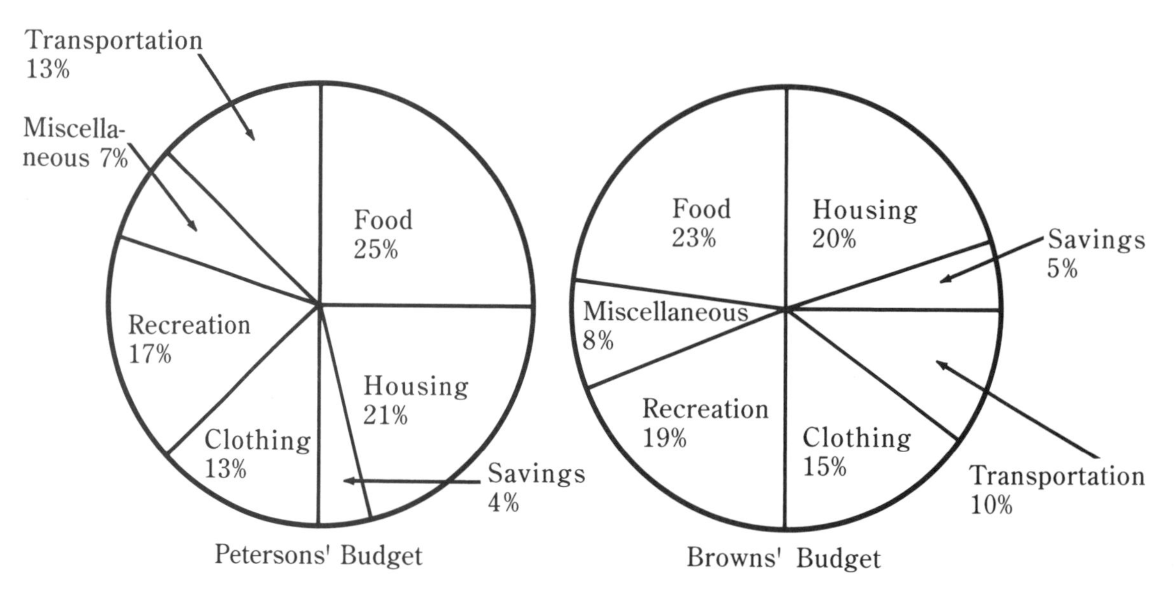

1. Which family spends more on food? What percent more?
2. Which family spends more on clothing? What percent more?
3. Which family saves more? What percent more?
4. On which two things do the Petersons spend the same percent of their income? What percent?
5. Which two things combined make up nearly 50% of both the Petersons' and Browns' budget?
6. Which family spends more on food, clothing, and recreation altogether? What percent more?
7. Which family spends more on transportation and recreation altogether? What percent more?
8. From the graph, can you tell how much money the Petersons spend on housing?

13.4 Using Your Calculator: Graphs and Percents

Circle graphs often give facts as percents of a number. If you know the number, you can use your calculator to find what numbers the percents stand for.

You have already learned how to use your calculator to find percents. If you need help, turn back to page 197 before completing the practice below.

PRACTICE

The city of Dawsonville collects 80,000 pounds of solid waste a month.

Study the graph below. Use your calculator to find the actual number of pounds that each section represents.

Solid Waste

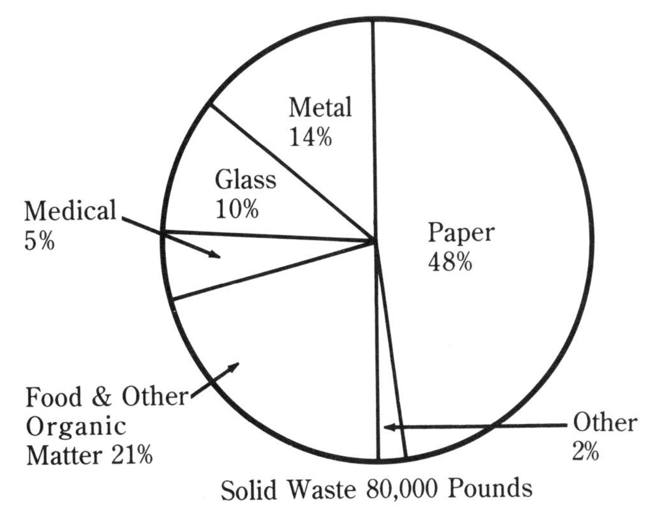

CHAPTER REVIEW

CHAPTER SUMMARY

- **Graph** — A graph is a picture that shows information as number facts.
- **Bar Graph** — Bar graphs use bars of different lengths to compare amounts or sizes of things.
- **Line Graph** — Line graphs use line segments to show changes and relationships between things over a period of time.
- **Circle Graph** — A circle graph is a graph in the shape of a circle. Circle graphs are used to show how a total amount has been divided into parts.

REVIEWING VOCABULARY

Copy the sentences on a separate sheet of paper. Put in the correct word or group of words from the box.

graph	line graph	bar graph	circle graph

1. A ________ is used to compare amounts or sizes of things.
2. A ________ is used to show changes over a period of time.
3. A ________ is used to show how a total amount has been divided.
4. Any ________ is a picture that shows information as number facts.

CHAPTER QUIZ

A. Study the circle graph. Answer the questions on a separate sheet of paper.

1. What is the percentage of students continuing their education?
2. Which group is the largest?
3. What percentage of students did not continue their education?

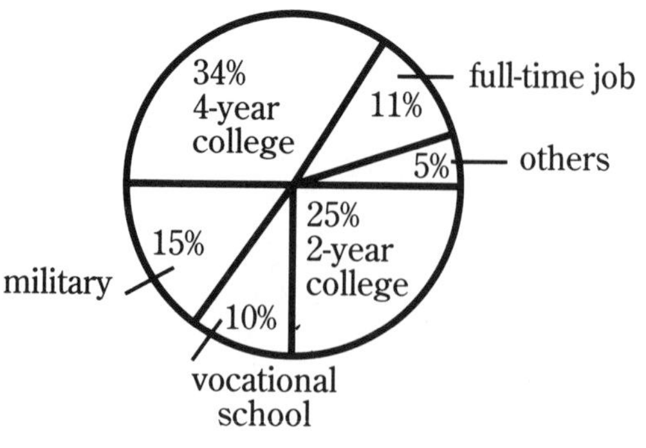

B. Study the graph below. Answer the questions on a separate sheet of paper.

Number of Employees

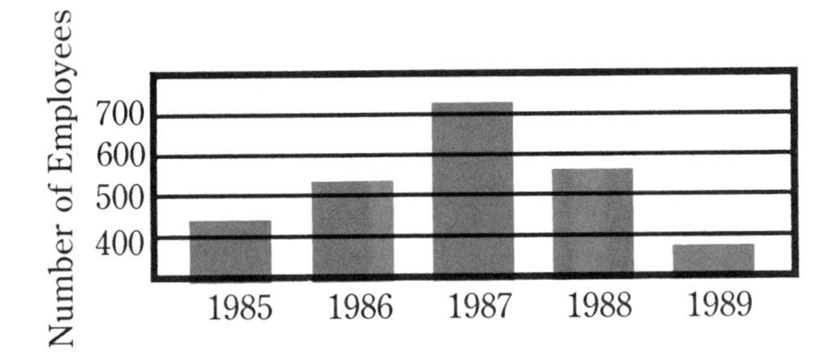

1. In what year were there the most employees? How many were there?
2. In what year were there the fewest employees? How many were there?
3. About how many more employees were there in 1987 than in 1989?

C. Study the graph below. Answer the questions on a separate sheet of paper.

Sales in Thousands of Dollars

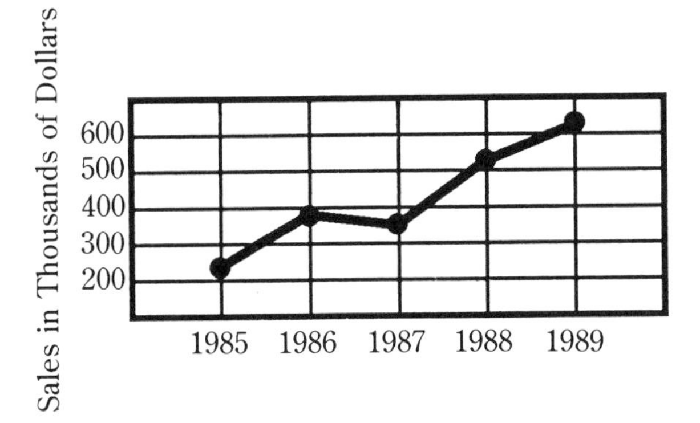

1. In what year were sales highest? About how much were sales?
2. In what year were sales lowest? About how much were sales?
3. Between which two years was there the greatest difference in sales? How much was this difference?

STATISTICS AND PROBABILITY

14

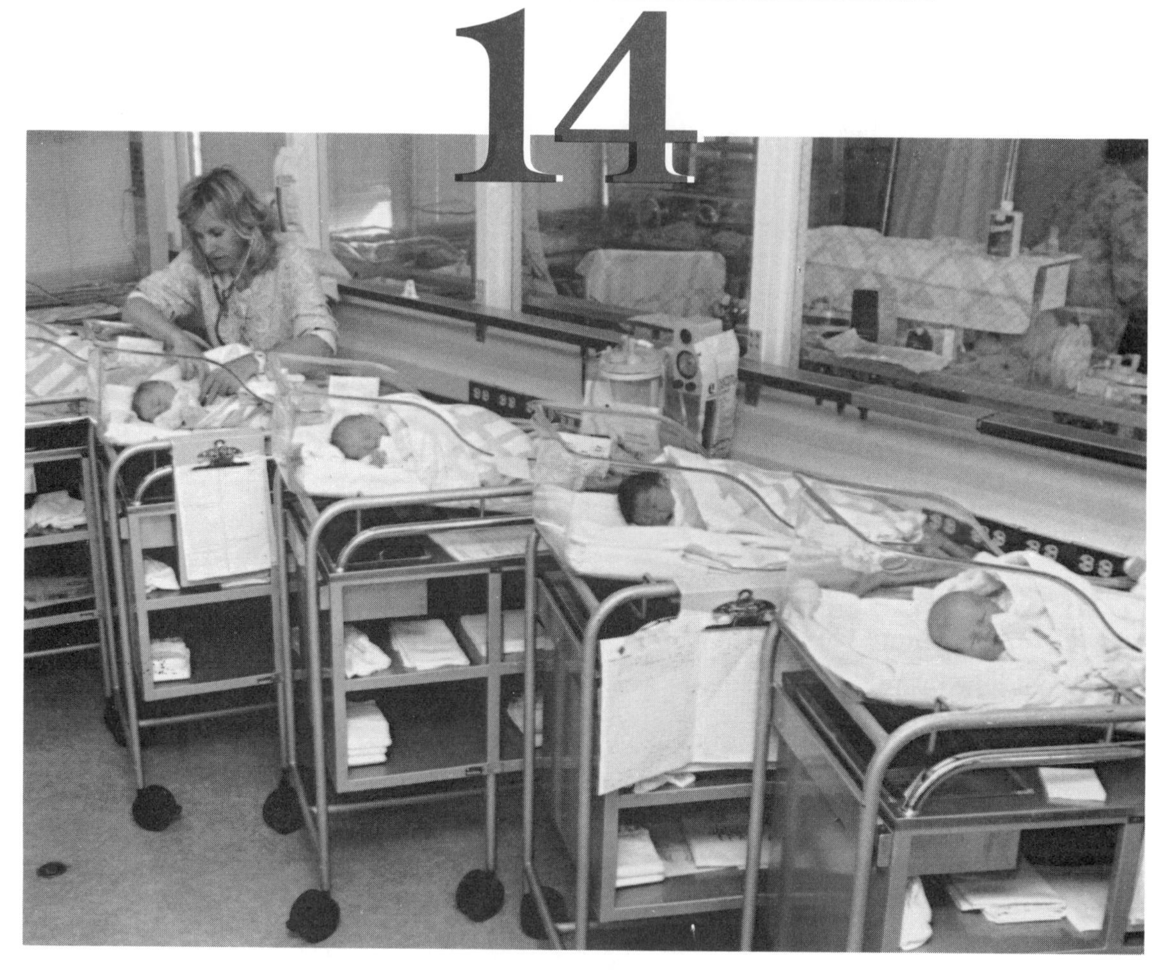

U.S. statistics showed that more than 3.5 million babies were born each year during the 1980s. The probability of a couple having a boy (or girl) is 1 chance in 2. In this chapter, you will learn all about statistics and probability.

Chapter Learning Objectives

1. Define and organize statistics
2. Find the mean or average of a set of numbers
3. Find the median of a set of numbers
4. Find the mode of a set of numbers
5. Find the range of a set of numbers
6. Find relative frequency in a set of numbers
7. Explain and demonstrate probability

WORDS TO KNOW

- **statistics** numerical information
- **mean** the number obtained by dividing the sum of a set of quantities by the number of quantities in the set; also called the average
- **median** the middle number in a set of numbers when the numbers are arranged in order of size
- **mode** the number in a set of numbers that occurs most frequently
- **range** the difference between the highest and lowest numbers in a set
- **frequency** the number of times a number or a range of numbers occurs in a set
- **probability** the likelihood that an event will occur

14.1 Gathering Statistics

A **statistic** is numerical information.

People who work with statistics do two jobs.

First they gather numerical information, and then they organize and explain that information.

Sometimes they also make decisions based on that information.

Some collections of statistics contain a great deal of varied information. The federal government, for example, takes a census every ten years. The census results in so much information that it takes the government several years to organize and print it.

A census is an official count of a country's population. A "census taker" also gathers statistics about the age, race, sex, and occupation of the population.

Most statistics are gathered on specific topics and for specific purposes. For example, a drug company develops a new drug. But before the drug can be sold, the company must test it on sick people. It must keep careful records, or gather statistics on how well the drug works. It must also gather statistics on any side effects the drug may have. Then the company must get statistics about sick people who did not take the drug. This second group is called a *control group*.

Surveys or polls provide statistics about people's opinions. Polls are taken before every election to see how people plan to vote. Of course poll takers cannot question everyone in the country. They can only talk to a small group, or a sample, of the population.

A *survey* is a collection of sample opinions. The information is used to form a general overall view of a situation. A *poll* is a kind of survey.

Polling a sample group can show how many people in general will vote. But it can do this only if the sample represents the people in general.

PRACTICE

Discuss these questions.

1. Why is a control group necessary when testing a new drug?
2. Imagine that a poll was taken in your neighborhood. The people chosen for the sample all live in the same apartment house. Do you think the poll will show the opinions of the whole neighborhood? Why or why not?

14.2 Frequency

The number of times a number or a range of numbers occurs in a set is called **frequency.**

Frequency helps you understand what a group of statistics means.

Dr. Ellis wanted to find out how much time the students at Washington High School spent watching television. She asked 60 students. Here are the numbers of hours each student watched television during one week.

8	12	22	9	17	15	19	25	32	26	7	18
13	25	27	5	28	18	41	35	27	29	9	21
28	38	26	17	32	43	36	23	33	26	11	18
29	40	16	9	30	19	25	44	20	10	26	37
42	39	27	22	16	31	28	26	19	22	39	20

The overall **range** of time these students watched television is from 5 hours to 44 hours. Dr. Ellis wanted to make these numbers easier to understand. So she grouped them into smaller ranges of hours. Each range contained 5 hours.

She put the numbers into a frequency table.

Range of hours	Tally	Frequency
5 to 9	卌 1	6
10 to 14	1111	4
15 to 19	卌 卌 1	11
20 to 24	卌 111	8
25 to 29	卌 卌 卌	15
30 to 34	卌	5
35 to 39	卌 1	6
40 to 44	卌	5

Dr. Ellis can make a graph to show this distribution. This kind of a graph is called a *histogram.* Notice that a histogram has no spaces between the bars.

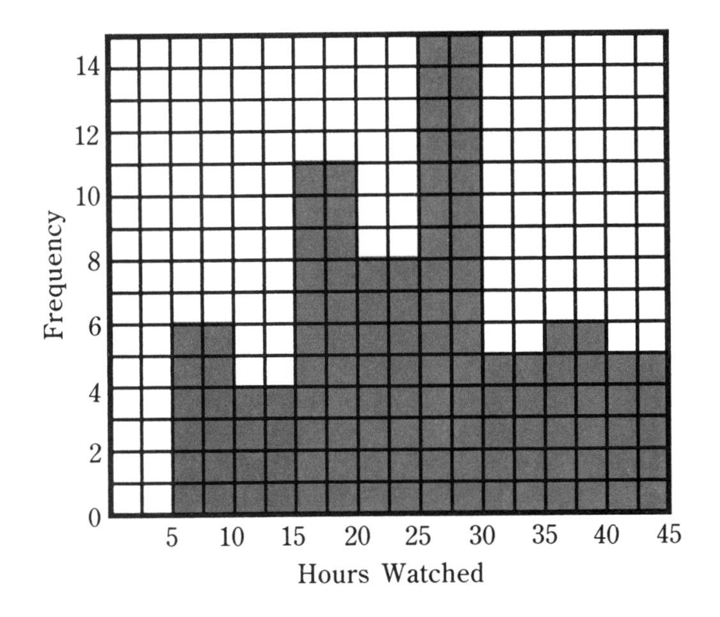

PRACTICE

Look at the group of statistics below. Divide them into five ranges of numbers: from $10.00 to $10.99, from $11.00 to $11.99, from $12.00 to $12.99, from $13.00 to $13.99, from $14.00 to $14.99. On a separate sheet of paper make a frequency table and a histogram for the numbers.

Amount Spent Weekly for Lunch

$10.75	$13.80	$12.75	$14.60	$11.95	$14.55
$11.25	$12.50	$14.55	$10.95	$12.45	$13.00
$13.20	$12.70	$12.15	$14.30	$13.36	$12.65
$12.15	$13.40	$12.60	$11.75	$12.00	$12.95

14.3 Relative Frequency Predictions

Poll takers often use the results of a sample survey to make predictions.

For instance, 125 students were asked if they had ever eaten in the school cafeteria. The results showed that 105 had eaten there, 20 had not.

To find the relative frequency of students who had never eaten in the cafeteria, follow the steps below.

Step 1: Write a fraction to show the total number of students and the number that hadn't eaten at the cafeteria.

$$\frac{20}{125}$$

Step 2: Multiply the fraction by 100%.

Remember that cancelling always makes multiplying fractions easier.

$$\frac{20}{\underset{5}{\cancel{125}}} \times \frac{\overset{4}{\cancel{100\%}}}{1} = \frac{80\%}{5} = 16\%$$

The relative frequency is 16%.

There are a total of 2,200 students in the school. You now have enough information to make a good guess or prediction. How many of these students are likely not to have eaten at the cafeteria? Just multiply the relative frequency by the total student population.

$$16\% = .16$$

$$\begin{array}{r} 2,200 \\ \times \quad .16 \\ \hline 13200 \\ 2200 \\ \hline 352.00 \end{array}$$

PRACTICE

Solve the problems below on a separate sheet of paper.

1. Sue checks cassettes before they leave the factory. Of the 168 she inspected, 2 were damaged. How many are likely to be damaged out of 4,400?

2. Stan surveyed 160 students at his high school. He wanted to know how much interest there was in forming a soccer team. He found 12 students who were interested. There are a total of 5,860 students at Stan's school. In all, how many are likely to be interested in soccer?

3. A telephone survey asked 324 people if they approved of the mayor's new traffic plan. Of those people questioned, 102 said they approved. The town's population is 72,038. How many people are likely to approve of the plan?

4. A car manufacturer asked 185 people if the owner's manual for their car was easy to understand. Of those questioned, 120 said it was not easy to understand. The manufacturer knows that 2,000,000 people own this kind of car. How many are likely to find the owner's manual difficult?

14.4 Mean

The **mean** is the number you get when you divide the sum of a set of quantities by the number of quantities in the set. The mean is also called the average.

Here is the formula for finding the mean:

$$\text{Mean} = \frac{\text{Sum of the quantities}}{\text{Number of quantities}}$$

Solve this problem. Follow the steps.

Six people used Ace batteries in their flashlights.

This list shows how long each pair of batteries lasted.

12 months
16 months
9 months
11 months
11 months
13 months

What is the mean life of a pair of Ace batteries?

Step 1: Find the sum of all the months.

$$\begin{array}{r} 12 \text{ months} \\ 16 \text{ months} \\ 9 \text{ months} \\ 11 \text{ months} \\ 11 \text{ months} \\ +13 \text{ months} \\ \hline 72 \text{ months} \end{array}$$

Step 2: Count the number of people.

6 people

Step 3: Divide the sum of the months by the number of people.

$$6\overline{)72}^{\,12}$$

The mean is the quotient you get.

The mean life of a pair of Ace batteries is 12 months.

PRACTICE

On a separate sheet of paper find the mean of each set of numbers.

1. 32, 24, 56, 85, 77, 49

2. 153, 208, 491, 622, 198

3. 461, 430, 500, 413, 479, 287

4. 234, 276, 209, 275, 288, 350, 525

5. 374, 682, 600, 550, 425

14.5 Median

The **median** is the middle number in a set of numbers arranged by size.

The numbers below are arranged from smallest to largest. Notice the number in the middle. It is the median.

3 5 8 *10* 14 16 19

Sometimes a set contains an even number of numbers. Then the middle of the set falls between two numbers.

Paul arranged the numbers about the life of batteries from the smallest number to the largest.

9 11 11 | 12 13 16

The median comes between 11 and 12.

To find the median, just find the average or mean of the middle two numbers.

Follow the steps below.

Step 1: Add the numbers on each side. $11 + 12 = 23$

Step 2: Divide the sum by 2. $23 \div 2 = 11\frac{1}{2}$

The median is $11\frac{1}{2}$, or 11.5.

PRACTICE

On a separate sheet of paper find the median in each set of numbers. Remember to arrange the numbers by size.

1. 55, 43, 17, 73, 29, 51

2. 171, 89, 138, 211, 146

3. 30, 20, 26, 24

4. 238, 276, 209, 235

5. 674, 592, 460, 630, 525

14.6 Mode

The **mode** is the number or range of numbers that occurs most frequently in a set. If each number or range of numbers occurs once, there is no mode.

Study this set of numbers. Notice the mode.

3 5 6 *8* *8* 9 11 13

The mode is 8.

Look again at Dr. Ellis's frequency table about television watching. Notice the mode. The mode is 25–29.

Range of Hours	Tally	Frequency
5 to 9	1111 1	6
10 to 14	1111	4
15 to 19	1111 1111 1	11
20 to 24	1111 111	8
25 to 29	1111 1111 1111	15
30 to 34	1111	5
35 to 39	1111 1	6
40 to 44	1111	5

You can use a frequency table to find the mode.

PRACTICE

On a separate sheet of paper find the mode in each set of numbers.

1. 5, 5, 4, 3, 1, 7, 7, 3, 2, 9, 5, 1

2. 31, 72, 89, 31, 38, 21, 31, 89, 72, 31, 46

3. 30, 20, 23, 27, 23, 26, 30, 23, 25, 30, 23, 26, 24

4. 209, 276, 207, 209, 235, 209, 238, 276, 209, 235

5. 525, 460, 465, 525, 674, 592, 525, 460, 630, 525, 630

14.7 Probability

Probability is the likelihood that something will happen.

For example: You toss a coin. What is the probability of the coin coming up heads?

This problem can be solved mathematically. There are only two ways a coin can land. One way is heads up. The second way is tails up. A coin has 1 chance in 2 of landing heads up. The possibility of a coin landing heads up is $\frac{1}{2}$ or 50%.

This kind of probability is called mathematical probability.

There is also another kind of probability. It is called experimental probability. To find this kind of probability, you must experiment.

Suppose a coin is old. It has worn unevenly. The chances of this coin coming up heads is not likely to be 50%. You would need to experiment to find the probability of this coin coming up heads.

Imagine this experiment. You toss the coin 100 times. It comes up heads 15 times. The experimental probability of the coin coming up heads is $\frac{15}{100}$ or 15%.

PRACTICE

A. Copy this chart on a separate sheet of paper. Toss an old coin 100 times. Use the chart to keep a record of how many times the coin comes up heads or tails. Then write the experimental probability of the coin coming up heads.

	Heads	Tails
First 25 times		
Second 25 times		
Third 25 times		
Fourth 25 times		
Total		

Experimental probability of the coin coming up heads: %

B. Read the problems. Choose one of the answers below each problem. Write your answers on a separate sheet of paper.

1. There are 18 marbles in a bag. Six of them are red. What is the probability of picking a red marble out of the bag the very first time?

20% 33% 50%

2. A regular deck of playing cards contains 52 cards. There are 13 clubs, 13 spades, 13 hearts, and 13 diamonds. What is the probability of pulling a club card from a full deck?

10% 20% 25%

3. Dave bought 10 tickets for a raffle prize. Exactly 500 tickets were sold. What is the probability that one of Dave's tickets will be chosen for the prize?

1% 2% 15%

MATHEMATICS IN YOUR LIFE:
Life Insurance

The table below shows how much someone would have to pay each year for each $2,000 worth of life insurance.

Two kinds of policies are offered. One is a straight life insurance policy. If the insured person dies, the beneficiary of the policy collects the amount of the insurance. The other policy is a 20-year endowment. That policy pays the amount of the insurance whether or not the insured person dies. In 20 years, either the beneficiary or the insured person can collect the insurance.

The amount you have to pay depends on:

—How old you are when you first buy the insurance
—Whether you are male or female
—What kind of insurance you want

Age	Straight Life		20-Year Endowment	
	Male	Female	Male	Female
18	23.98	22.10	83.96	83.60
19	24.62	22.70	84.08	83.24
20	25.34	23.32	84.16	83.84
21	26.04	23.98	84.24	83.96
22	26.78	24.62	84.32	84.08
23	27.58	25.34	84.40	84.16
24	28.42	26.04	84.48	84.32
25	29.28	26.78	84.54	84.40

Use this table to solve the problems on the next page.

Solve this problem. Follow the steps.

Carl is 22 years old. He wants to buy a $20,000 straight life insurance policy. How much will he have to pay each year?

Step 1: Find out how much $2,000 worth of straight life insurance would cost each year.

$26.78

Step 2: Divide the total amount of insurance by 2,000.

$$2{,}000\overline{)20{,}000} = 10$$

Step 3: Multiply the yearly payment by the quotient.

$$\$26.78 \times 10 = \$267.80$$

Solve the problems below on a separate sheet of paper.

1. Janice is 19 years old. She wants to buy a $14,000, 20-year endowment policy. How much will she have to pay each year?
2. A 24-year-old man and a 24-year-old woman both want to buy $38,000 20-year endowment policies. How much more will the man have to pay for the policy each year than the woman?
3. Zack and Rob are both 20 years old. Each of them wants to buy $10,000 worth of insurance. Zack chooses a straight life policy. But Rob decides that a 20-year endowment policy is best for him. Which policy costs more per year? How much more?
4. Mary carries four times as much straight life insurance as her brother Mark does. Mary is 23 and Mark is 22. Mary pays $101.36 a year for her insurance. How many thousands of dollars of insurance does Mary carry? How much does Mark carry? How much does Mark pay a year for his insurance?
5. Max wants to buy a $50,000 straight life policy. How much will he have to pay over 5 years for this policy? What do you need to know first before you can solve this problem?

14.8 Using Your Calculator: Finding the Mean

You can use your calculator to find the mean.

Solve this problem. Find the mean of these numbers: 486, 594, 327.

Step 1: Press [4] [8] [6] [+]

Step 2: Press [5] [9] [4] [+]

Step 3: Press [3] [2] [7] [=]

Step 4: Press [÷]

Step 5: Press [3] [=]

Step 6: Your answer should be: 469

Use your calculator to find the mean of each group of numbers below. Round your answer to the nearest whole number.

1. 7,432, 5,924, 5,626, 8,435, 2,477, 4,309
2. 1,653, 3,208, 4,891, 1,622, 2,198
3. 4,561, 4,330, 5,700, 4,913, 4,079, 7,612
4. 6,234, 5,276, 5,209, 6,275, 5,288
5. 9,374, 8,682, 8,600, 7,550, 7,425, 9,432
6. 4,625, 6,409, 753, 38,502
7. 8,626, 5,907, 4,333, 7,207
8. 8,995, 2,666, 1,085, 9,950
9. 5,025, 8,006, 7,322, 10,505
10. 4,690, 7,038, 6,502, 5,756

CHAPTER REVIEW

CHAPTER SUMMARY

- **Statistics** Statistics give information in the form of numbers. Statistics can be gathered through surveys and polls. Statistics can be used to predict what will happen.
- **Mean** The mean is obtained by dividing the sum of a set of quantities by the number of quantities in the set. The mean is sometimes called the average.
- **Median** The median is the middle number in a set of numbers arranged by size. When there is an even number of numbers in a set, the median is the average of the middle two numbers.
- **Mode** The mode is the number in a set of numbers that occurs most frequently. If each number occurs once, there is no mode.
- **Range** The range is the difference between the highest and lowest numbers in a set.
- **Frequency** Frequency is the number of times a number or range of numbers occurs.
- **Probability** Probability is the likelihood that an event will occur.

REVIEWING VOCABULARY

Number a separate sheet of paper from 1 to 7. Read each word in the column on the left. Find its definition in the column on the right. Write the letter of the definition next to each number.

1. frequency
2. mode
3. probability
4. range
5. statistics
6. mean
7. median

a. numerical information
b. average
c. the middle number of a set of numbers when the numbers are arranged in order of size
d. the number in a set of numbers that occurs most frequently
e. the difference between the highest and lowest numbers in a set
f. the number of times a number or a range of numbers occurs
g. the likelihood an event will occur

CHAPTER REVIEW

CHAPTER QUIZ

A. Reviewing Statistics and Probability Skills

Read this group of numbers:

773 629 842 794 773 698

1. Find the range of the group of numbers above.

2. Find the median of the group of numbers above.

3. Find the mode of the group of numbers above.

4. Find the mean of the group of numbers above.

5. Make a frequency table for this group of numbers.

35 32 37 29 37 32 30 32 37 36 37 35
27 29 27 35 37 30 37 35 37 27 32 37

6. Use the frequency table you made in problem 5 to find the mode.

7. Make a frequency table for these ranges of numbers.

25–34 35–44 15–24 35–44 25–34 35–44
1–14 15–24 35–44 1–14 35–44 15–24

8. Use the frequency table you made in problem 7 to find the mode.

B. Solve the following word problem on a separate sheet of paper.

Merle found that 22 out of a group of 78 students at City College drove to school. There are 3,486 students at the college. How many of them are likely to drive to school?

C. Developing Thinking Skills

David works on commission. He made $2,542 the first month. He made $3,142 the second month and $2,973 the third month. He wants to estimate how much he is likely to make the fourth month. Should he find the median or the mean? Why do you think so?

D. Using the table on page 237, find the annual cost of a $10,000 straight life policy for a male who is twenty-three.

UNIT FOUR REVIEW

Study the graph below. Answer the questions on a separate sheet of paper.

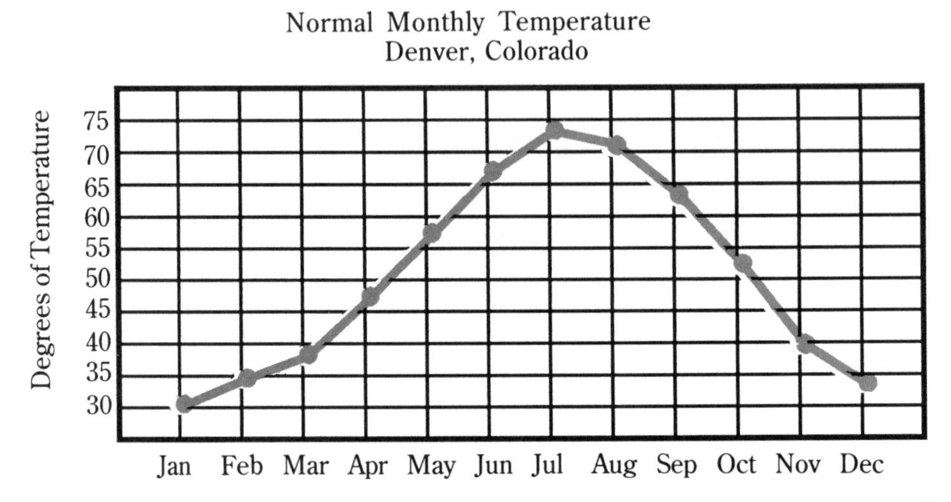

1. During which month does Denver have its *lowest* normal temperature? What is the temperature?
2. During which month does Denver have its *highest* normal temperature? What is the temperature?
3. What is the *approximate* difference between Denver's highest and lowest temperature throughout the year?
4. *Approximately* what is the single biggest *drop* in temperature from one month to the next?

Solve the problems below on a separate sheet of paper.

Read this group of numbers:

864 518 953 682 891 587

5. Find the range of the group of numbers above.
6. Find the median of the group of numbers above.
7. Find the mode of the group of numbers above.
8. Find the mean of the group of numbers above.

UNIT FIVE

15
CUSTOMARY MEASUREMENT

16
METRIC MEASUREMENT

CUSTOMARY MEASUREMENT

15

This Chicago snowstorm dumped more than 10 inches of snow on the city. At what temperature does rain turn to snow? Is the answer 0°—or 32°? Both answers are correct. To find out how that is possible, read this chapter.

Chapter Learning Objectives

1. Use customary measurements
2. Convert customary measurements to larger and smaller units
3. Measure temperature
4. Measure perimeter
5. Measure area
6. Measure volume
7. Find distance, rate, and time

WORDS TO KNOW

- **customary** something that is done or based on accepted practice; usual or commonly used
- **convert** to change into a different form; to exchange for something of equivalent value
- **perimeter** the distance or boundary around a shape
- **area** the surface a shape covers
- **volume** the space inside a container; volume is also called capacity
- **rate** a measured amount in relation to a fixed quantity of something else; a *rate* of speed is how fast something is going
- **Fahrenheit** a scale of measuring temperature
- **Celsius** a scale of measuring temperature

15.1 Measurements

Measurements give length, width, area, weight, time, and speed.

Below are some units of **customary** measurement. Customary means usual, or commonly used.

The word customary comes from the word *custom*. Can you think of any customs we follow today that were followed hundreds of years ago?

Time

60 seconds = 1 minute
60 minutes = 1 hour
24 hours = 1 day

7 days = 1 week
52 weeks = 1 year
12 months = 1 year

Length

12 inches = 1 foot
36 inches = 1 yard

3 feet = 1 yard
5,280 feet = 1 mile

Weight

16 ounces = 1 pound

2,000 pounds = 1 ton

Liquid Volume

16 fluid ounces = 1 pint
2 pints = 1 quart

32 fluid ounces = 1 quart
4 quarts = 1 gallon

PRACTICE

Copy this list on a separate sheet of paper. Next to each item write the unit of customary measurement it is measured by.

1. your height
2. length of a knife
3. coal in freight car
4. your weight
5. length of a room
6. distance between two cities
7. weight of a bag of flour
8. milk in a bottle
9. weight of a hot dog
10. time worked in a day

15.2 Changing Measurements to Smaller Units

It is easy to **convert** a larger unit of measurement to a smaller one. You can do this by multiplying.

You have already *converted* things in an earlier chapter. Look back at Chapter 7, lessons 3 and 4. Notice how you had to convert fractions to their highest and lowest terms.

Change 6 feet to inches.

You know that there are 12 inches in 1 foot.

Multiply 6 feet by 12 inches.

$6 \times 12 = 72$

There are 72 inches in 6 feet.

Change $3\frac{1}{2}$ hours to minutes.

You know that there are 60 minutes in 1 hour.

Multiply $3\frac{1}{2}$ hours by 60 minutes.

$$3\frac{1}{2} \times 60 =$$

$$\frac{7}{\cancel{2}_1} \times \frac{\cancel{60}^{30}}{1} = \frac{210}{1} = 210$$

There are 210 minutes in $3\frac{1}{2}$ hours.

PRACTICE

Change these units of measure on a separate sheet of paper.

1. Change 9 yards to feet.
2. Change 3 tons to pounds.
3. Change 5 pounds to ounces.
4. Change 4 yards to inches.
5. Change 8 weeks to days.
6. Change 13 hours to minutes.
7. Change 7 quarts to pints.
8. Change 6 pints to ounces.
9. Change 11 gallons to quarts.
10. Change 3 miles to feet.

15.3 Changing Measurements to Larger Units

It's also possible to convert a smaller unit of measurement to a larger one. You can do this by dividing.

Change 108 inches to feet.

You know that there are 12 inches in 1 foot.

Divide 108 inches by 12 inches.

$108 \div 12 = 9$

There are 9 feet in 108 inches.

Change 80 minutes to hours.

You know that there are 60 minutes in 1 hour.

Divide 80 minutes by 60 minutes.

$$\begin{array}{r} 1 \text{ R } 20 \\ 60\overline{)\,80} \\ \underline{60} \\ 20 \end{array} \quad \text{or } 1\tfrac{20}{60} = 1\tfrac{1}{3} \text{ hours}$$

PRACTICE

Change these measures on a separate sheet of paper.

1. Change 36 feet to yards.
2. Change 5,325 pounds to tons.
3. Change 27 ounces to pounds.
4. Change 57 inches to yards.
5. Change 45 days to weeks.
6. Change 204 minutes to hours.
7. Change 83 pints to quarts.
8. Change 35 ounces to pints.
9. Change 41 quarts to gallons.
10. Change 12,010 feet to miles.

15.4 Measuring Temperature

Temperature can be measured in two ways. You can use the **Fahrenheit** scale, or you can use the **Celsius** scale.

The unit of measure for the Fahrenheit scale is degrees Fahrenheit, written °F. The unit of measure for the Celsius scale is degrees Celsius, written °C.

Study the difference in the way the two scales measure the same temperature.

	Fahrenheit	**Celsius**
freezing point of water	32°F	0°C
boiling point of water	212°F	100°C
normal body temperature	98.6°F	37°C
lowest recorded air temperature	−127°F	−88°C
highest recorded air temperature	136°F	58°C

Notice the lowest recorded air temperatures. They have minus signs before them.

A minus sign means that the temperature is that many degrees below 0.

Converting Temperatures

It is possible to convert a temperature on one scale to the same temperature on the other.

To change a Celsius temperature to a Fahrenheit reading, use this method:

$$F^\circ = \left(\frac{9}{5} \times C^\circ\right) + 32$$

For example: Change 100°C to the same temperature on the Fahrenheit scale.

Step 1: $F^\circ = \left(\frac{9}{5} \times 100^\circ\right) + 32$

Step 2: $\frac{9}{\cancel{5}_1} \times \frac{\cancel{100^\circ}^{20^\circ}}{1} = \frac{180^\circ}{1} = 180^\circ$

Step 3: $180^\circ + 32 = 212^\circ$

$212^\circ F = 100^\circ C$

To convert a Fahrenheit temperature to a Celsius reading use this method:

$$C^\circ = \frac{5}{9} \times (F^\circ - 32)$$

For example: Change 77°F to the same temperature on the Celsius scale.

Step 1: $C^\circ = \frac{5}{9} \times (77^\circ - 32)$

Step 2: $77^\circ - 32 = 45^\circ$

Step 3: $\frac{5}{9} \times 45^\circ$

$\frac{5}{\cancel{9}_1} \times \frac{\cancel{45^\circ}^{5}}{1} = \frac{25^\circ}{1} = 25^\circ$

$77^\circ F = 25^\circ C$

PRACTICE

Number a separate sheet of paper from 1 to 9. Change each temperature below to an equal temperature on the other scale.

1. 0°C
2. 41°F
3. 68°F
4. 35°C
5. 15°C
6. 95°F
7. 55°C
8. 113°F
9. 40°C

15.5 Finding the Perimeter

The **perimeter** is the distance or boundary around a shape or area.

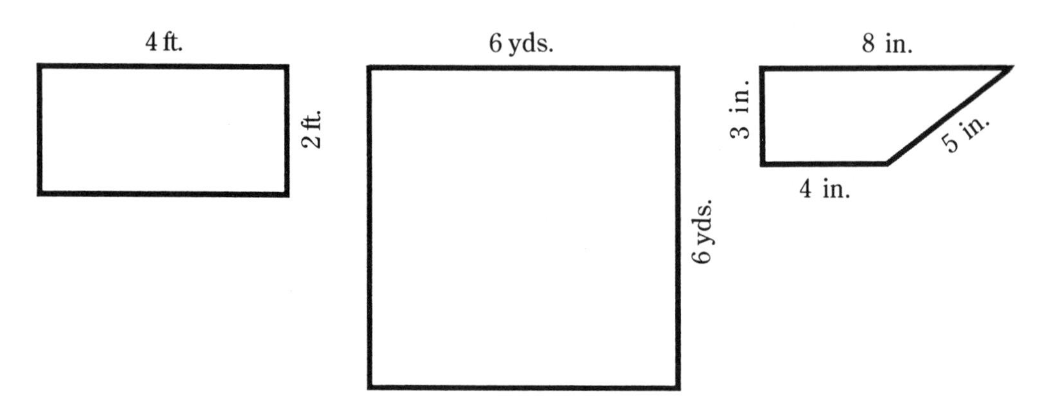

You can find the perimeter of a shape by adding the lengths of the sides.

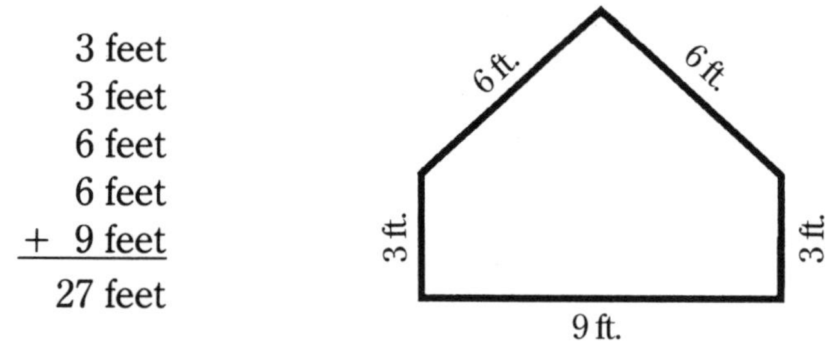

PRACTICE

Find the perimeter of each shape. Work on a separate sheet of paper.

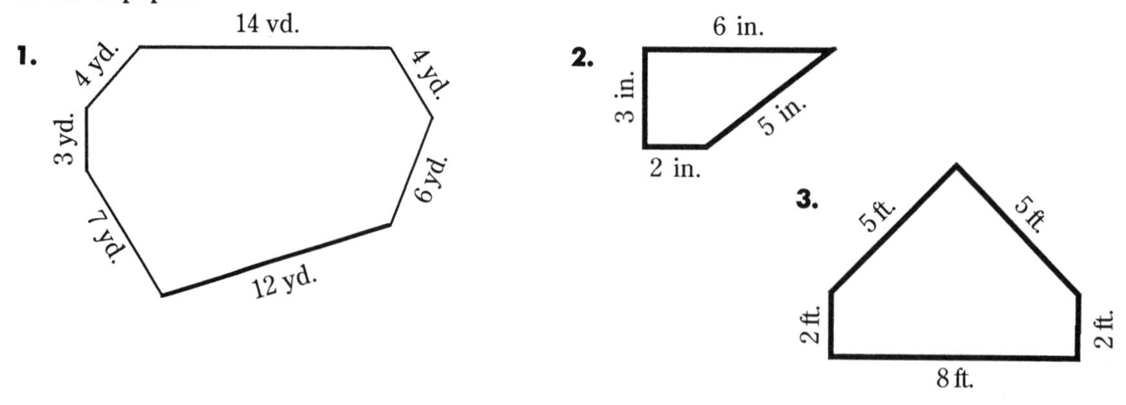

Many shapes have four sides. Sometimes their opposite sides are equal. You can find the perimeter of shapes like these. Look at the shape below. Follow the steps.

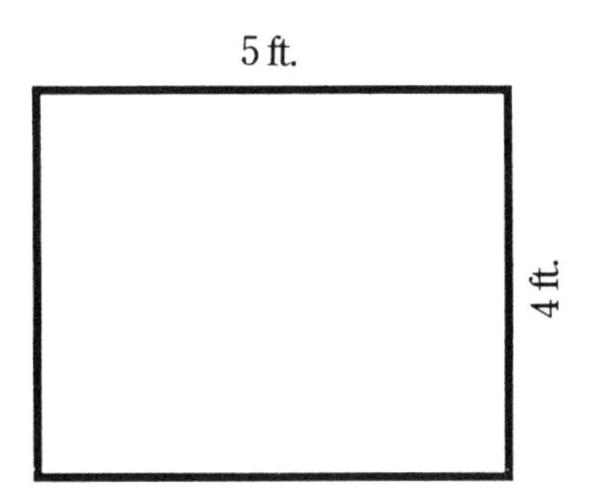

Step 1: Multiply the length by 2. (There are 2 lengths.) $5 \times 2 = 10$

Step 2: Multiply the width by 2. (There are 2 widths.) $4 \times 2 = 8$

Step 3: Add the sums. $10 + 8 = 18$

The perimeter of this shape is 18 feet.

Perimeter $= 2 \times$ length $+ 2 \times$ width

PRACTICE

On a separate sheet of paper, find the perimeters of these shapes.

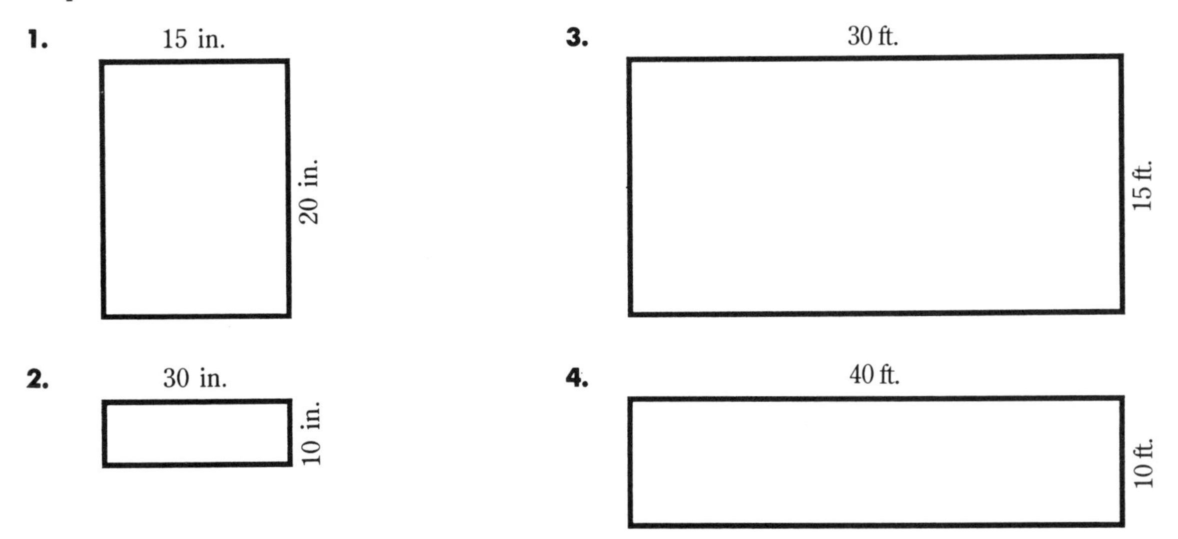

15.6 Square Inches, Feet, and Yards

Look at the pictures.

1 in.

1 ft.

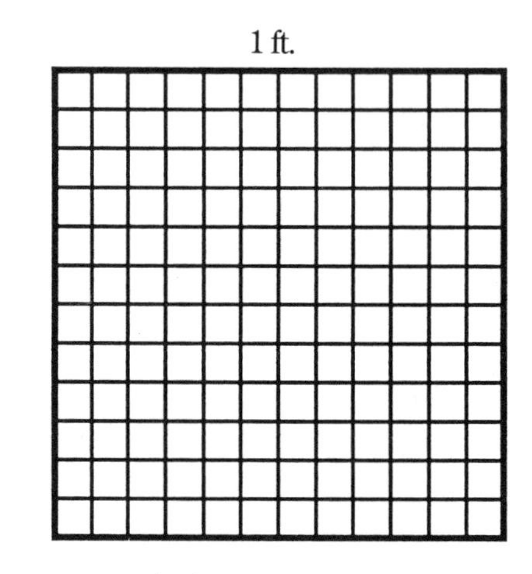

The first square stands for an area that is one inch long and one inch wide. This area is called one square inch. The second square stands for a space that is one foot long and one foot wide. This area is called one square foot. It is divided into square inches. Count the number of square inches in one square foot. A square that is one yard long and one yard wide is called one square yard.

PRACTICE

Count the number of square yards in each shape. Write the answers on a separate sheet of paper.

3 yd.

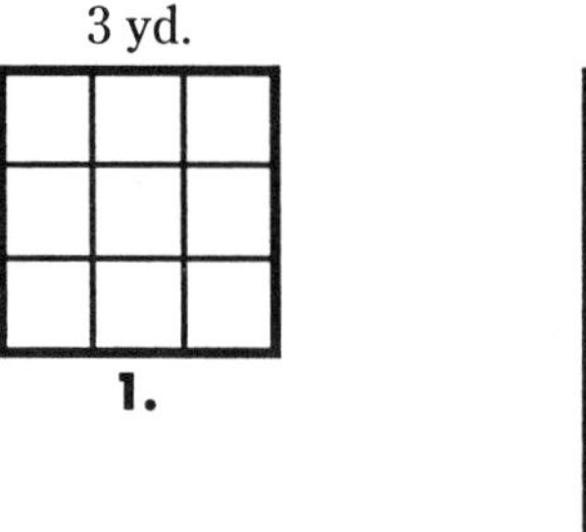

1.

5 yd.

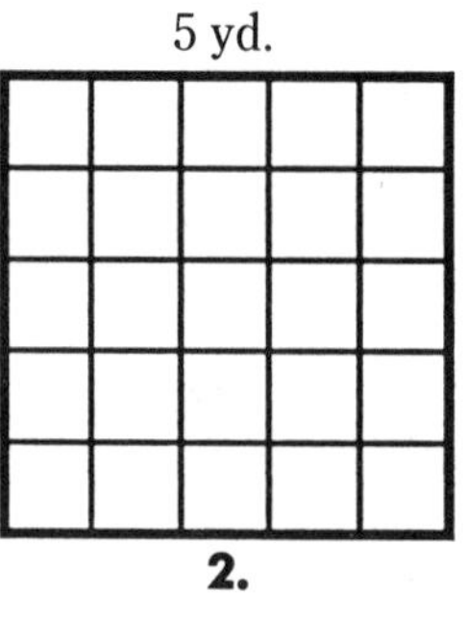

2.

6 yd.

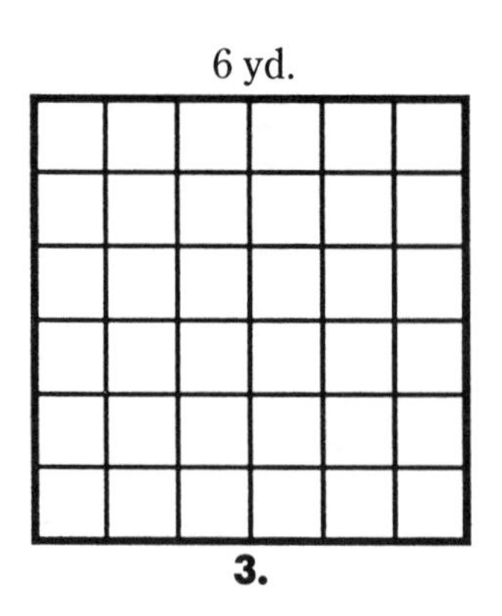

3.

15.7 Finding the Area

The **area** is the amount of surface a shape covers.

Areas are measured in square units such as square inches, square feet, square yards, and square miles.

You can find the area of a shape or a place if you know its length and width. But remember the length and width have to be written in the same units.

Follow the steps below.

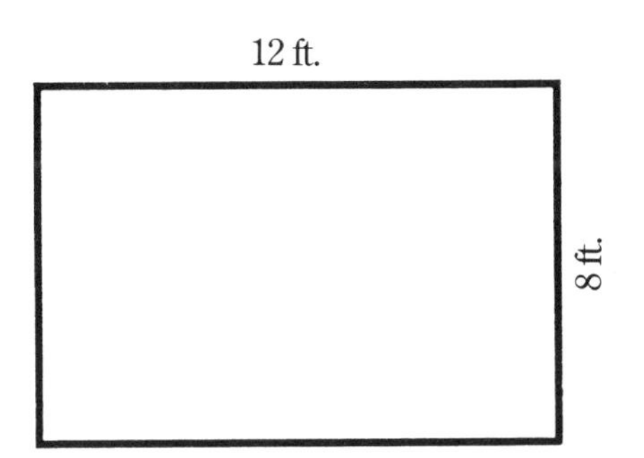

Step 1: Find the length and the width.

Step 2: Multiply the length by the width.

Step 3: Label your answer as square feet.

length = 12 ft
width = 8 ft
12 × 8 = 96
12 × 8 = 96 sq ft

The area of this shape is 96 sq ft.

Area = length × width

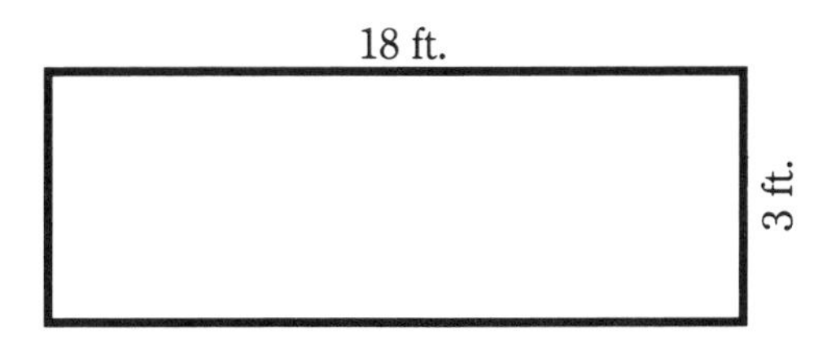

length = 18 ft
width = 3 ft
18 × 3 = 54
18 × 3 = 54 sq ft
The area of this shape is 54 sq ft.

PRACTICE

Find the areas of each shape.

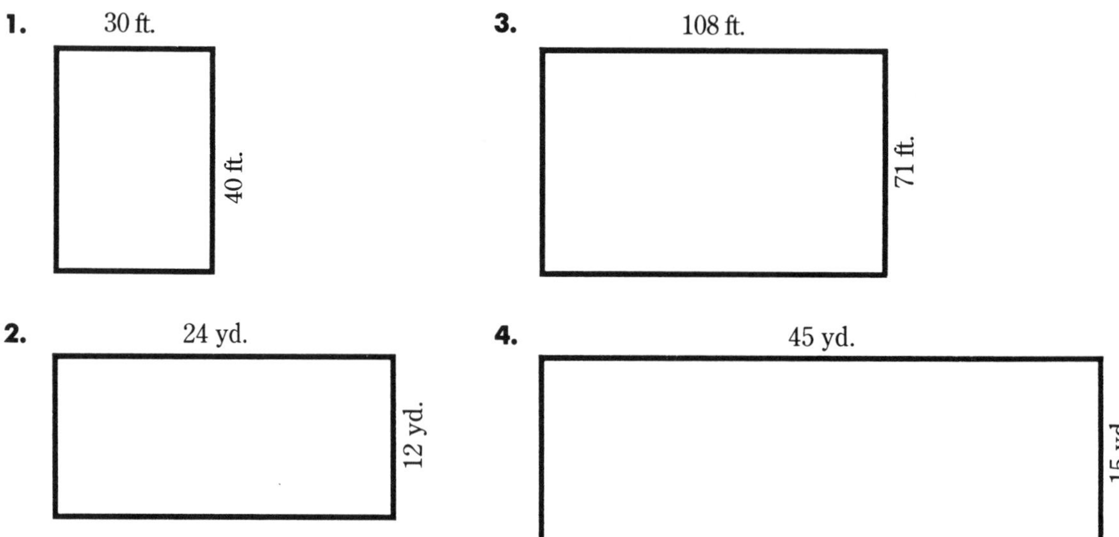

15.8 Finding Volume

Volume measures the space inside a container.

Look at the picture. Notice the measurements for length, width, and depth.

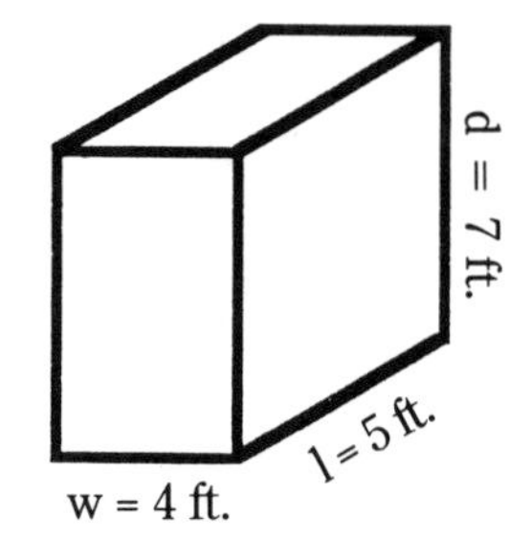

Volume (or capacity) is measured in cubic units such as cubic inches, cubic feet, and cubic yards.

A cubic measure has three dimensions: length, width and depth. This is the reason volumes are measured in cubic units.

Find the volume of a container.

Step 1: Find the length, width, and height.

Step 2: Multiply the length by the width.

Step 3: Multiply this product by the height.

Step 4: Label your answer as cubic feet.

The volume of this container is 702 cubic ft.

Volume = length × width × height

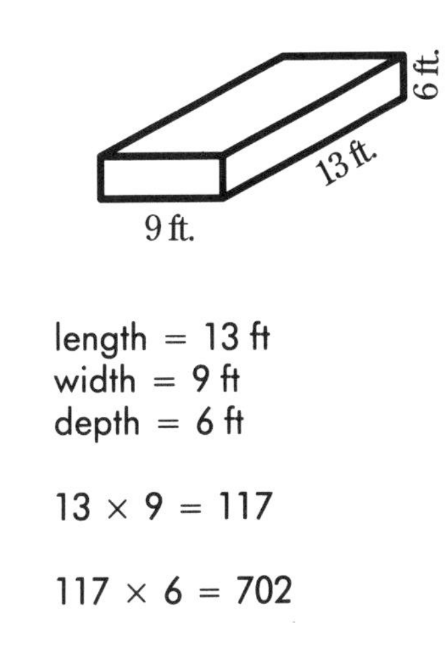

length = 13 ft
width = 9 ft
depth = 6 ft

$13 \times 9 = 117$

$117 \times 6 = 702$

702 cu ft

PRACTICE

Find the volume of each container.

1.

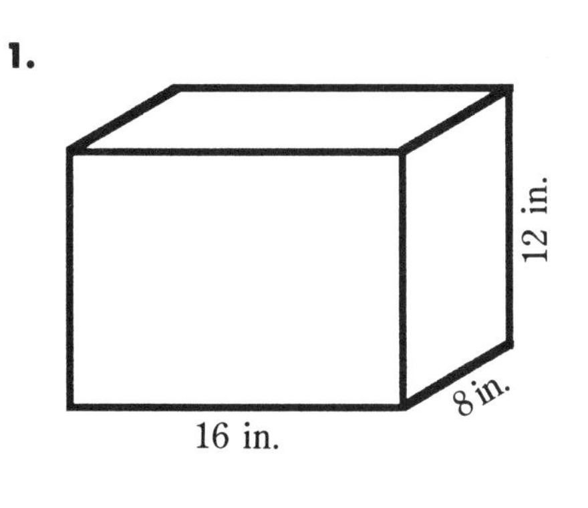

2.

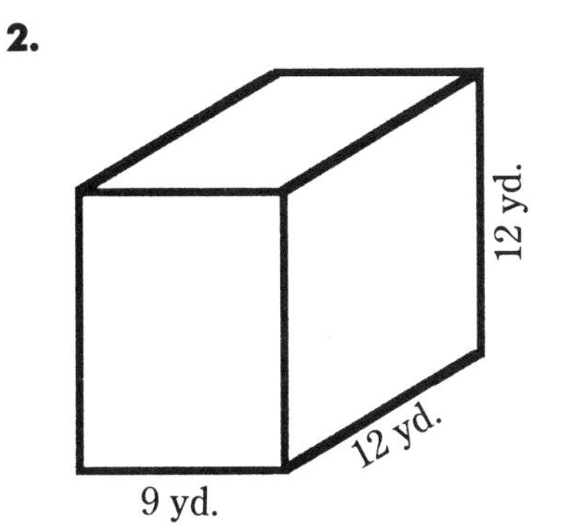

3.

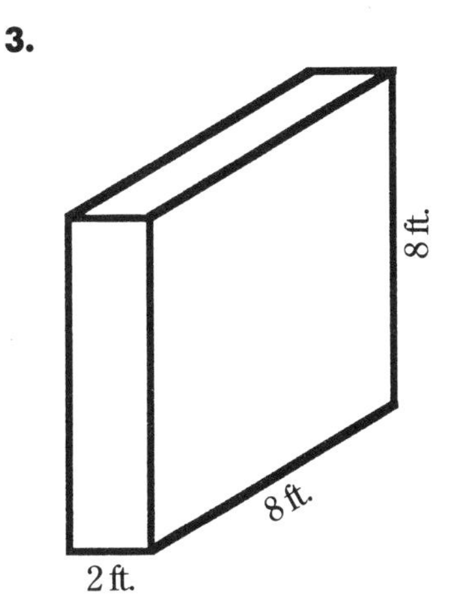

4.

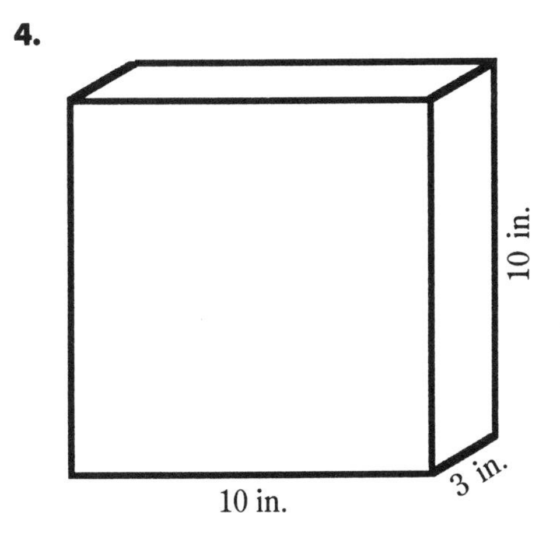

15.9 Distance, Time, and Rate

It is easy to determine the distance something has traveled if you know two things:

1. the **rate** of speed that it is traveling
2. the length of time it has traveled

Rate $\times$ Time $=$ Distance

$R \times T = D$

For example:

Jenny drove for 5 hours at 55 miles an hour. How far did she drive?

$R \times T = D$

$55 \times 5 = 275$

Jenny drove 275 miles.

If you know the distance and time, you can find out the rate.

$$\frac{\text{Distance}}{\text{Time}} = \text{Rate}$$

Harry drove 78 miles in 2 hours. At what rate per hour did he drive?

$\frac{78}{2} = 39$

Harry drove at a rate of 39 miles an hour.

If you know the distance and rate, you can find out the time.

$$\frac{\text{Distance}}{\text{Rate}} = \text{Time}$$

A plane flew 960 miles. Its rate of speed was 320 miles an hour. How long did the trip take?

$\frac{960}{320} = 3$

The trip took 3 hours.

PRACTICE

Solve the problems below on a separate sheet of paper.

1. Gerald drove for 6 hours at 47 miles an hour. How far did he go?
2. A plane went 945 miles in 3 hours. What was the plane's average hourly rate?
3. A train went 350 miles at 75 miles an hour. How long did the trip take?
4. Jack drove for 4 hours at 52 miles an hour. Tim drove for 5 hours at 43 miles an hour. Who drove farther? How much farther did he drive?

15.10 Solving Measurement Problems

Solve the measurement problem below. Follow the steps.

The Clinton family used 2 gallons of milk one week. They used 1 gallon and 3 quarts the next week. How many quarts did they use in all?

Step 1: Write the numbers as an addition problem.

	2 gallons	
+	1 gallon	3 quarts
	3 gallons	3 quarts

Step 2: Find the sum.

Step 3: Change the gallons in the sum to quarts. (Remember there are 4 quarts in a gallon.)

$3 \times 4 = 12$

Step 4: Add the quarts.

$12 + 3 = 15$

The Clinton family used 15 quarts of milk altogether.

PROBLEMS TO SOLVE

Solve the problems below on a separate sheet of paper.

1. Cory used 3 gallons of paint on Monday. On Tuesday he used 2 gallons and 1 quart. On Wednesday he used 3 quarts. How many quarts did he use altogether?

2. Harry worked 21 hours and 45 minutes the first week. He worked 22 hours and 15 minutes the second week. How many hours did he work in all?
3. A pool is 120 feet long. Marlene wants to swim 1 mile. How many times does she have to swim the length of the pool?

MATHEMATICS IN YOUR LIFE: Homemade Recipes

Rita and Carlos are selling homemade tomato sauce.

"We can make two gallons of sauce in this pot," said Carlos. "I wonder how many quart jars that much sauce will fill?"

"There are four quarts in a gallon," said Rita. "With two gallons we can fill eight jars."

1. Suppose Rita and Carlos decided to put the sauce in pint jars. How many jars would they need for 3 gallons of sauce?
2. Carlos and Rita decide to put 5 gallons of sauce in pint jars. They also decide to put 4 gallons of sauce in quart jars. How many pint jars do they need? How many quart jars do they need?
3. Copy this chart on a separate sheet of paper. Complete the chart. Show how many items of food are needed to make the different amounts of sauce.

	1 gallon	3 gallons	5 gallons
tomatoes	6 pounds		
onions	2 pounds		
celery		3 bunches	
peppers	$1\frac{1}{2}$ pounds		
parsley			$1\frac{1}{2}$ bunches

15.11 Using Your Calculator: Measurements

You can use your calculator to figure amounts and measurements. Solve this problem. Follow the steps.

Bob's Service Station is open 24 hours a day, 7 days a week. How many hours is it open in a year?

Step 1: Press [2] [4]

Step 2: Press [×]

Step 3: Press [7]

Step 4: Press [×]

Step 5: Press [5] [2]

Step 6: Press [=]

Step 7: Your answer should be: 8,736 hours

PRACTICE

1. Barbara worked 6 hours a day, 4 days a week for 49 weeks. How many hours did she work in all?
2. In Wilderness Park one hiking trail is 12 miles long. Another is 17 miles long. A third is 9 miles long. The Park Service wants to double the length of the trails. If they do that how many miles of hiking trails would there be altogether?
3. Linda is 5 feet 3 inches tall. What is her height in inches?
4. Susan's first paycheck was for 13 days. Her second paycheck was for 18 days. Her third paycheck was for 11 days. How many weeks did these three paychecks cover?

CHAPTER REVIEW

CHAPTER SUMMARY

- **Customary Measurement** In the United States, people usually use the customary measurement system for everyday measurements. Terms such as *inch, foot, mile, ounce,* and *pound* are used in the customary measurement system.
- **Temperature** Temperature may be measured with either the Fahrenheit or the Celsius scale.
- **Perimeter** The perimeter is the distance around a shape. Perimeter may be measured by adding the lengths of all the sides. The perimeter of certain four-sided shapes may be found by multiplication and addition: 2 × length + 2 × width (2 × l + 2 × w).
- **Area** Area is the surface a shape covers. Area may be found by multiplying length × width (l × w).
- **Volume** Volume is the space inside a container. Volume may be found by multiplication: length × width × height (l × w × h).
- **Distance** Distance = Rate × Time; D = R × T.

REVIEWING VOCABULARY

Copy the sentences on a separate sheet of paper. Put the correct word from the box in each blank.

perimeter	area	volume	Fahrenheit	Celsius

1. ________ is the space inside a container.

2. ________ is the distance around a shape.

3. ________ and ________ are ways of measuring temperature.

4. ________ is the surface a shape covers.

CHAPTER REVIEW

CHAPTER QUIZ

A. Solve the problems below. Write your answers on a separate sheet of paper.

1. 4 feet = ________ inches.

2. 32 ounces = ________ pounds.

3. A room is 16 feet long and 12 feet wide. What is its perimeter?

4. A field is 27 yards long and 35 yards wide. What is its area?

5. A box is 17 inches long, 13 inches wide, and 3 inches high. What is its volume?

6. Helen ran for 2 hours. She went 7 miles. How many miles did she run per hour?

B. Change the following temperatures to equal temperatures on the other scale.

1. 65°C　　**2.** 25°C　　**3.** 59°F　　**4.** 86°F

C. Developing Thinking Skills

Study the shapes below. Then answer the questions that follow.

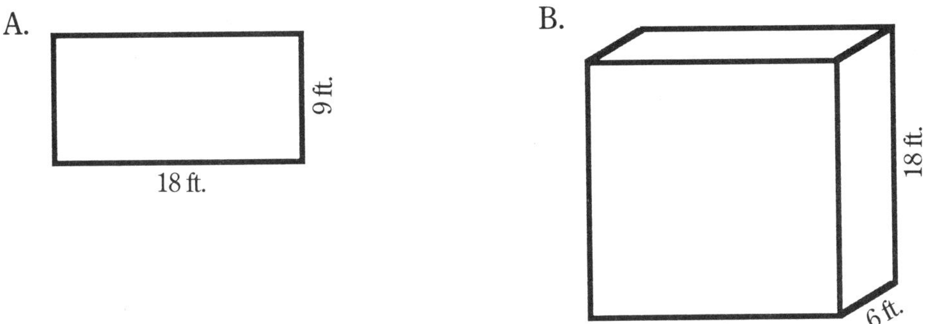

1. What is the length of Shape A? What is its width? What is the shape's perimeter? What is its area? What is its perimeter in *yards?* What is its area in *yards?*

2. What is the length of Shape B? What is its width? What is its height? What is the volume of the shape in *inches?* What is the volume of the shape in *yards?*

Metric Measurement

16

This Olympic weightlifter is raising a weight of 97.5 kilograms. That is nearly 215 pounds. All Olympic events are scored on the metric system of measurement.

Chapter Learning Objectives

1. Identify the vocabulary of the metric system
2. Compare amounts in the metric and customary systems
3. Measure length with the metric system
4. Measure weight with the metric system
5. Measure volume with the metric system

WORDS TO KNOW

metric system	a system of measurement based on the number 10
meter	a metric unit of measurement used to measure *length*
gram	a metric unit of measurement used to measure *weight*
liter	a metric unit of measurement used to measure *capacity*
kilo-	a prefix that stands for 1,000; a *kilometer* is 1,000 meters
hecto-	a prefix that stands for 100; a *hectogram* is 100 grams
deka-	a prefix that stands for 10; a *dekaliter* is 10 liters
deci-	a prefix that stands for .1 (one-tenth); *a decimeter* is one-tenth of a meter
centi-	a prefix that stands for .01 (one-hundredth); a *centigram* is one-hundredth of a gram
milli-	a prefix that stands for .001 (one-thousandth); a *milliliter* is one-thousandth of a liter

16.1 What Is the Metric System?

In the last chapter you learned about customary measurement. Now you will learn about a different system of measurement called the **metric system.**

Metric measurement is used in most countries around the world. It is an easy system to use because it is based on the number ten.

You already know how to use systems based on ten. Our number system is based on ten. You use money all the time. Our money system is also based on 10.

Study the examples on the next page.

Notice that ten of a smaller coin equals one of a larger coin. You will also notice that ten of a smaller bill equals one of a larger bill.

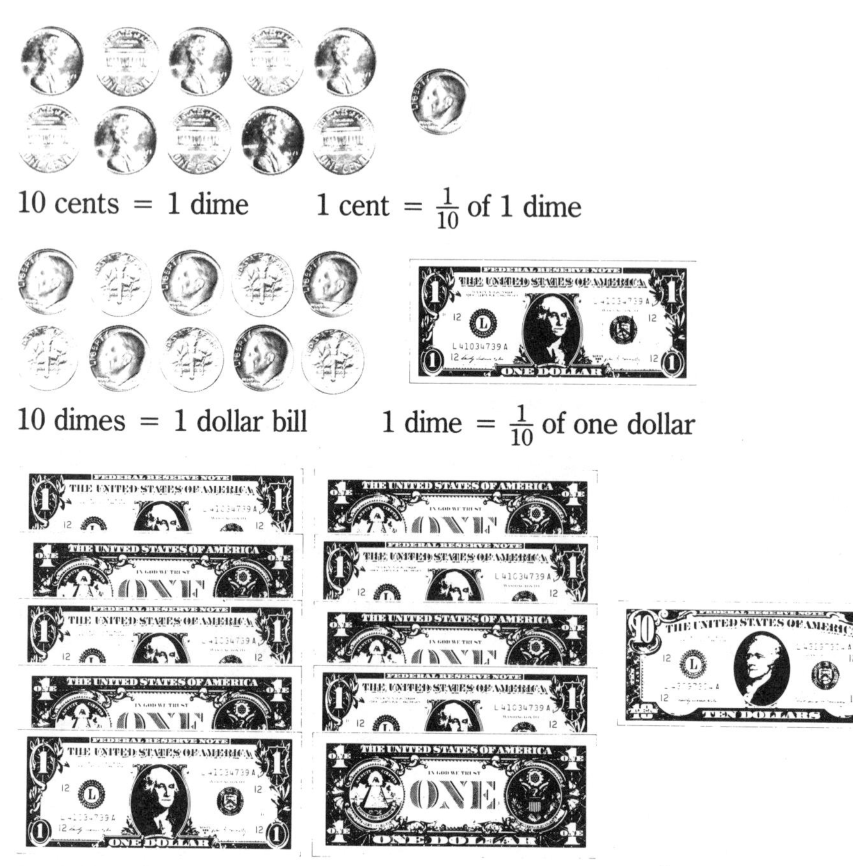

10 cents = 1 dime   1 cent = $\frac{1}{10}$ of 1 dime

10 dimes = 1 dollar bill   1 dime = $\frac{1}{10}$ of one dollar

10 dollar bills = 1 ten dollar bill   1 dollar bill = $\frac{1}{10}$ of 1 ten dollar bill

PRACTICE

Copy these sentences on a separate sheet of paper. Use the pictures of money above to help you fill in the blanks.

1. One 10 dollar bill equals ________ 1 dollar bills.
2. One 1 dollar bill equals ________ of one 10 dollar bill.
3. One 1 dollar bill equals ________ dimes.
4. One cent equals ________ of one dime.
5. One 100 dollar bill equals ________ 10 dollar bills.
6. One 10 dollar bill equals ________ of one 100 dollar bill.
7. One 10 dollar bill equals ________ dimes.

16.2 Comparing Metric and Customary Measurements

Study the pictures below. Compare the size of the two measurements in each one.

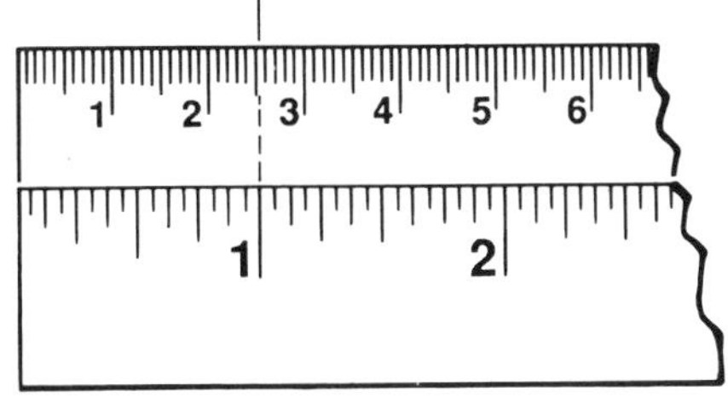

About 2.54 centimeters = 1 inch

1 liter = about 2.1 pints

About 1.6 kilometers = 1 mile

About 28.4 grams = 1 ounce

PRACTICE

Complete the number sentences below on a separate sheet of paper.

1. 1 ounce = about ________ grams

2. 1 liter = about ________ pints

3. 1 mile = about ________ kilometers

4. 1 inch = about ________ centimeters

5. 5 liters = about ________ pints

6. 12 miles = about ________ kilometers

7. 6 ounces = about ________ grams

8. 8 inches = about ________ centimeters

16.3 Measuring Length

In the metric system, the **meter** is used to measure length. The abbreviation for meter is m.

All other measures of length are based on the meter.

An *abbreviation* is a shorter form of a word. It is made up of the first or first few letters of the word followed by a period. What other abbreviations can you think of?

More than 1 meter	1 kilometer = 1,000 meters 1 hectometer = 100 meters 1 dekameter = 10 meters
	1 meter
Less than 1 meter	100 centimeters = 1 meter 10 millimeters = 1 centimeter 1,000 millimeters = 1 meter

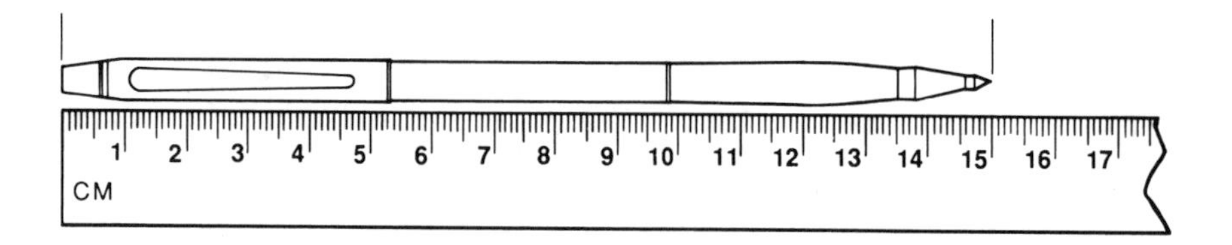

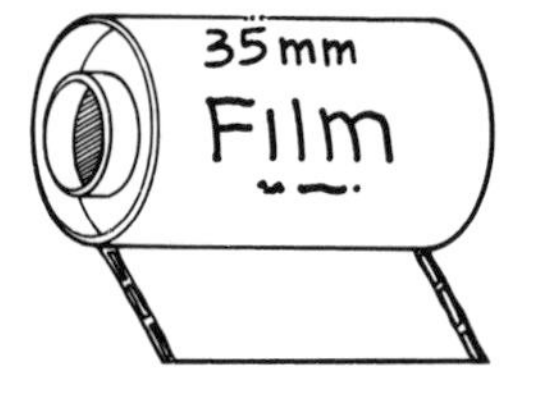

In the metric system it's easy to change a larger unit of length to a smaller one. All you have to do is multiply.

Change 6 dekameters to meters.

You know that there are 10 meters in 1 dekameter.

Multiply 6 dekameters by 10 meters.

$6 \times 10 = 60$

There are 60 meters in 6 dekameters.

It's also easy to change a smaller metric unit of length to a larger one. You just divide.

Change 208 centimeters to meters.

You know that there are 100 centimeters in 1 meter.

Divide 208 centimeters by 100.

$208 \div 100 = 2$ R8

There are 2 meters and 8 centimeters in 208 centimeters.

PRACTICE

Change these measurements on a separate sheet of paper.

1. Change 7 meters to centimeters.
2. Change 532 centimeters to meters.
3. Change 2,789 millimeters to meters.
4. Change 5 kilometers to meters.
5. Change 4 dekameters to meters.
6. Change 18 meters to millimeters.
7. Change 830 hectometers to meters.
8. Change 35,756 meters to kilometers.
9. Change 250 hectometers to meters.
10. Change 12,010 meters to kilometers.
11. Change 640 centimeters to millimeters.
12. Change 10 dekameters to centimeters.
13. Change 4,500 millimeters to meters.
14. Change 8 kilometers to hectometers.
15. Change 500 meters to dekameters.
16. Change 6,000 meters to kilometers.
17. Change 12 meters to centimeters.
18. Change 90 millimeters to centimeters.

Did You Know?
The blue whale is the world's largest animal. It may grow up to 100 feet (30 meters) long and can weigh as much as 200,000 pounds (91 metric tons). On an ordinary day, the blue whale can eat 400 pounds of fish.

16.4 Measuring Weight

In the metric system, the **gram** is used to measure weight. The abbreviation for gram is g.

All other measures of weight are based on the gram. One of the most common measures is the kilogram. The sign for kilogram is kg.

More than 1 gram	1 kilogram = 1,000 grams 1 hectogram = 100 grams 1 dekagram = 10 grams
	1 gram
Less than 1 gram	100 centigrams = 1 gram 10 milligrams = 1 centigram 1,000 milligrams = 1 gram

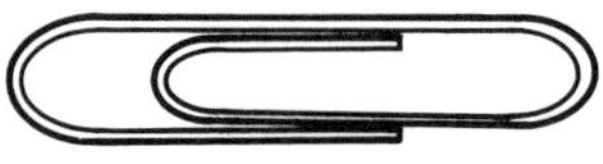
Weight = about 1 gram

Weight = about 1 kilogram

Changing larger metric units of weight to smaller ones is the same as changing units of length. You multiply.

Change 9 dekagrams to grams.

You know that there are 10 grams in 1 dekagram.

Multiply 9 dekagrams by 10 grams.

$9 \times 10 = 90$

There are 90 grams in 9 dekagrams.

The method for changing smaller units of weight to larger ones is dividing.

Change 340 centigrams to grams.

You know that there are 100 centigrams in 1 gram.

Divide 340 centigrams by 100.

$340 \div 100 = 3 \text{ R}40$

There are 3 grams and 40 centigrams in 340 centigrams.

PRACTICE

Change these measurements on a separate sheet of paper.

1. Change 8 kilograms to grams.
2. Change 674 centigrams to grams.
3. Change 3,572 milligrams to grams.
4. Change 92 kilograms to grams.
5. Change 7 dekagrams to grams.
6. Change 18 grams to milligrams.
7. Change 954 hectograms to grams.
8. Change 35,756 grams to kilograms.
9. Change 67,034 milligrams to grams.
10. Change 46 dekagrams to grams.
11. Change 8,400 centigrams to grams.
12. Change 18 kilograms to hectograms.
13. Change 16,240 milligrams to grams.
14. Change 42 grams to dekagrams.
15. Change 550 dekagrams to centigrams.
16. Change 50,000 grams to kilograms.
17. Change 12,500 grams to hectograms.
18. Change 550 milligrams to centigrams.

16.5 Measuring Volume

In the last chapter you learned that volume or capacity is the space inside a container. In the metric system, the liter is used to measure volume or capacity. The abbreviation for **liter** is l.

All other measures of volume are based on the liter.

More than 1 liter	1 kiloliter = 1,000 liters 1 hectoliter = 100 liters 1 dekaliter = 10 liters
	1 liter
Less than 1 liter	100 centiliters = 1 liter 10 milliliters = 1 centiliter 1,000 milliliters = 1 liter

Holds about 1 milliliter

Holds about 1 liter

Changing larger metric units of capacity to smaller ones is simple. Again, all you need do is multiply.

Change 4.5 dekaliters to liters.

You know that there are 10 liters in 1 dekaliter.

Multiply 4.5 dekaliters by 10 liters.

$4.5 \times 10 = 45$

There are 45 liters in 4.5 dekaliters.

If you need help multiplying by decimals, turn back to page 168.

To change smaller units of capacity to larger ones, you divide.

Change 475 centiliters to liters.

You know that there are 100 centiliters in 1 liter.

Divide 475 centiliters by 100.

$475 \div 100 = 4 \text{ R}75$

There are 4 liters and 75 centiliters in 475 centiliters.

PRACTICE

Change these measurements on a separate sheet of paper.

1. Change 9 liters to centiliters.
2. Change 895 centiliters to liters.
3. Change 5,629 milliliters to liters.
4. Change 7 hectoliters to liters.
5. Change 26 dekaliters to liters.
6. Change 32 liters to milliliters.
7. Change 830 hectoliters to liters.
8. Change 427 liters to hectoliters.
9. Change 731 hectoliters to liters.
10. Change 5,738 liters to dekaliters.
11. Change 420 dekaliters to hectoliters.
12. Change 1,200 milliliters to centiliters.
13. Change 950 centiliters to liters.
14. Change 1,570 liters to dekaliters.
15. Change 140 hectoliters to liters.
16. Change 53 kiloliters to liters.
17. Change 276 liters to centiliters.
18. Change 13,000 centiliters to milliliters.

16.6 Solving Metric Measurement Problems

Solve this measurement problem. Follow the steps below.

Susan used 20 liters of gas in one week. She used 25 liters the next week. The third week, she used 22 liters. In all, she drove 2,825 kilometers. How many kilometers does she average per liter?

Step 1: Write the numbers as an addition problem to find how many liters Susan used.

```
  20 liters
  25 liters
+ 22 liters
-----------
  67 liters
```

Step 2: Divide the number of kilometers she drove by the number of liters of gas she used. Take the quotient to two decimal places.

```
      42.16
67)2825.00
   268
   ---
    145
    134
    ---
     11 0
      6 7
     ----
      4 30
      4 02
      ----
        28
```

Susan averages 42.16 kilometers per liter.

PROBLEMS TO SOLVE

1. Kevin drove 145 kilometers on Monday. On Tuesday he drove 210 kilometers. On Wednesday he drove 189 kilometers. He drove 95 kilometers on Thursday and 179 kilometers on Friday. How many kilometers did he average each day? Find your answer to one decimal place.

2. A length of rope is 9 meters and 6 centimeters long. Another length is 12 meters and 3 centimeters long. How much longer is the second length of rope? Give your answer *in centimeters*.

3. Marlene bought 2 kilograms of flour. She used 250 grams on Saturday. On Sunday she used 24 grams. She used 360 grams on Tuesday and 700 grams on Thursday. How many grams were left? Did she have more or less than 1 kilogram left?

4. Mickey gathered 78 kilograms of apples from his trees. He divided them among himself and his two brothers. How many kilograms did each person get? How many *grams* did each one get?

5. A kitchen sink holds 20 liters of water. A pail holds 8 liters. How many pails of water would it take to fill up the sink? How many liters would be left in the pail?

6. A dog weighs 10 kilograms. A cat weighs 2 kilograms. How much do they weigh together? Give your answer in *grams*.

7. One bottle of cough medicine holds $\frac{1}{2}$ liter of medicine and one dropper full equals one milliliter. How many droppers will empty the bottle?

8. One paper clip weighs one gram. If four rolls of quarters are placed on a scale and weigh one kilogram, how many paper clips would be needed to balance the scale?

9. A ball-point pen is 15 centimeters long. A postage stamp is 15 millimeters long. How many stamps would you need to equal the length of one ball-point pen?

10. Don ran in the Los Angeles Marathon in March. The marathon is about 25.6 miles. How many kilometers is this? (Round your answer to the nearest one.)

MATHEMATICS IN YOUR LIFE: Distances

On most road signs in the United States, distances are given in miles. Only rarely are distances given in kilometers. But in most other countries, distances are given in kilometers.

Gerry and Laura are planning a trip. They plan to drive to several towns, starting from Asherville. The map below shows their route.

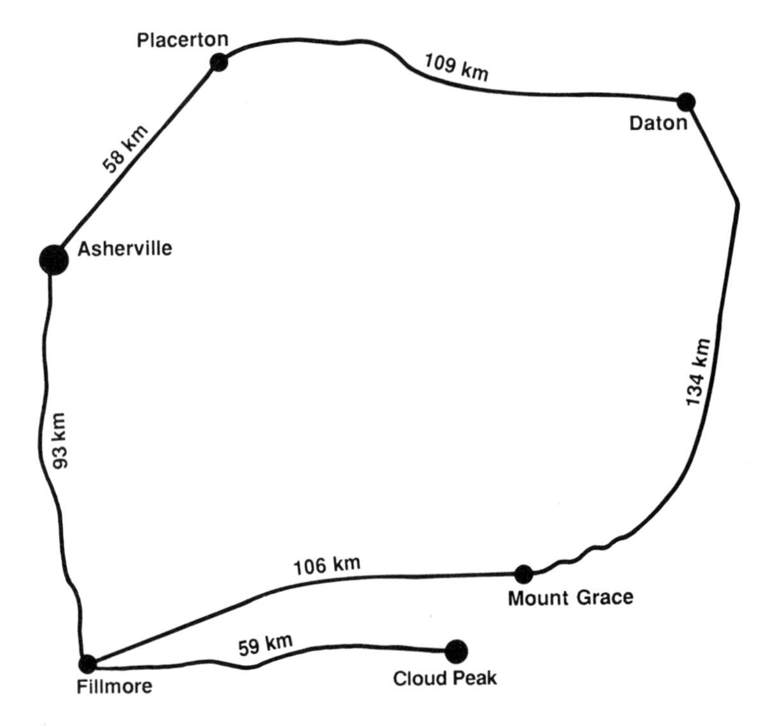

1. Between which two towns is the distance longest?
2. Between which two towns is the distance shortest?
3. How far will Gerry and Laura drive in all?
4. Gerry and Laura each plan to drive for half of the trip. How far will each drive?
5. Suppose Gerry and Laura decide to take a side trip to Cloud Peak when they leave Fillmore. How many kilometers will that add to their trip?

16.7 Using Your Calculator: Changing Measurements

You can use your calculator to convert customary measurements to metric measurements.

For example: change 14 inches into centimeters. You know that 1 inch = 2.54 centimeters. Just multiply the number of inches by 2.54.

$14 \times 2.54 = 35.56$ centimeters

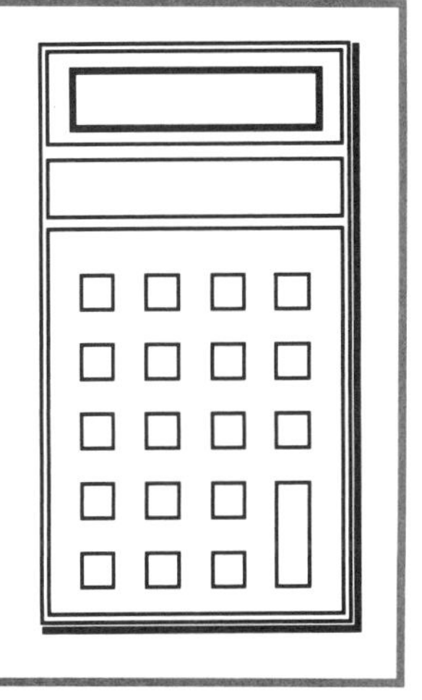

Study the chart below. It gives you the conversions from customary to metric measurements.

inches × 2.54 = centimeters	pints × .4732 = liters
feet × 30.48 = centimeters	quarts × .9464 = liters
yards × .9144 = meters	gallons × 3.785 = liters
miles × 1.609 = kilometers	pounds × .4536 = kilograms

PRACTICE

Use your calculator to convert these measurements.

1. Change 18 feet to centimeters.
2. Change 14 gallons to liters.
3. Change 32 miles to kilometers.
4. Change 23 miles to kilometers.
5. Change 16 quarts to liters.
6. Change 17 yards to meters.
7. Change 4 feet to centimeters.
8. Change 22 pounds to kilograms.
9. Change 13 gallons to liters.
10. Change 26 pints to liters.
11. Change 20 gallons to liters.
12. Change 53 feet to centimeters.

Chapter Review

CHAPTER SUMMARY

- **Metric system** — The metric system of measurement is based on ten.
- **Meter** — The meter is the basic unit for measuring length or distance. All other units for measuring length are based on the meter.
- **Gram** — The gram is the basic unit for measuring weight. All other units for measuring weight are based on the gram.
- **Liter** — The liter is the basic unit for measuring volume or capacity. All other units for measuring volume or capacity are based on the liter.

REVIEWING VOCABULARY

Copy the sentences on a separate sheet of paper. Put the correct word from the box in each blank.

kilometer	dekaliters	centigrams
metric	milliliters	

1. A ____________ is the same as 1,000 meters.
2. If you weighed a pen and pencil, the metric weight would be measured in ____________ .
3. The capacities of several quarts of milk can be measured in ____________.
4. Another system of measurement besides customary is the ____________ system.
5. Ten bottles of cola would most likely equal two ____________ .

CHAPTER REVIEW

CHAPTER QUIZ

A. Change the measurements below on a separate sheet of paper.

1. Change 8 liters to centiliters.
2. Change 6,549 meters to kilometers.
3. Change 4,635 grams to kilograms.
4. Change 9 hectoliters to liters.
5. Change 3 kilometers to meters.
6. Change 5,830 milligrams to grams.
7. Change 483 deciliters to liters.
8. Change 639 centigrams to grams.

B. Choose the likely measurements for each of the numbered sentences below. Write your answers on a separate sheet of paper.

1. The distance between two towns
   80 meters   80 kilometers   80 deciliters
2. The length of your living room
   16 dekameters   16 dekagrams   16 meters
3. The volume of a container of milk
   2 liters   2 grams   2 centimeters
4. The weight of a sack of potatoes
   20 kiloliters   20 kilograms   20 hectometers
5. The weight of a house key
   4 grams   4 kilograms   4 liters
6. The volume of a bathtub full of water
   5 hectograms   5 dekaliters   5 liters
7. The length of a man's belt
   1 meter   1 centimeter   1 dekameter
8. The weight of this textbook
   2 grams   2 liters   2 kilograms

UNIT FIVE REVIEW

Solve the problems below on a separate sheet of paper.

1. 5 quarts = ____ pints

2. 24 feet = ____ yards

3. 8 pounds = ____ ounces

4. 7 gallons = ____ pints

5. 9 liters = ____ centiliters

6. 4 kilometers = ____ meters

7. 5,620 grams = ____ kilograms

8. 12 hectoliters = ____ liters

9. 10 kilograms = ____ grams

10. 72 inches = ____ yards

11. Les wants to put a border around his living room. The room is 18 feet by 24 feet. How many feet of border does he need?

12. Amy measured her room for carpeting. It's 15 feet long and 9 feet wide. Carpet is sold by the square yard. How many square yards of carpet are needed to cover the room?

13. Phil and Laura built a garage. It was 50 feet by 95 feet. What was the area of the floor?

14. A carton is 72 inches long, 48 inches wide, and 84 inches high. What is its volume?

15. A plane flew for 6 hours at 532 miles an hour. How many miles did it fly?

16. Nora drove 509 miles in 9 hours. What was her rate of speed?

17. A box is 56 inches long, 38 inches wide, and 48 inches high. A second box is 68 inches long, 40 inches wide, and 3 inches high. Which box has the greater volume? How much greater is the volume of one box than the other?

Convert each temperature below to the other scale.

18. 75°C **19.** 104°F **20.** 45°C

UNIT SIX

17
POLYGONS

18
CIRCLES

19
EQUATIONS

POLYGONS

17

This is the most famous polygon-shaped building in our country—the U.S. Pentagon. It's called that because it has five sides. Pentagons are just one of the many polygons you'll learn about in this chapter.

Chapter Learning Objectives

1. Define polygon
2. Measure angles
3. Identify shapes such as quadrilaterals, rectangles, squares, parallelograms, and triangles
4. Find the perimeter of various polygons
5. Find the area of various polygons
6. Find the square roots of numbers

WORDS TO KNOW

■ **plane figure**	a figure that has only length and width
■ **polygon**	a plane figure with three or more sides
■ **angle**	the space between two straight lines meeting at a point
■ **degree**	the unit used to measure angles; the symbol for degree is °
■ **quadrilateral**	a plane figure with four sides
■ **rectangle**	a plane figure with four sides and four ninety degree angles; the opposite sides of a rectangle are equal
■ **square**	a plane figure with four equal sides and four ninety degree angles
■ **parallel lines**	lines that extend in the same direction and at the same distance apart; parallel lines can never meet
■ **parallelogram**	a plane figure with four sides; the opposite sides of a parallelogram are parallel
■ **triangle**	a plane figure with three sides
■ **formula**	a mathematical rule showing the relationship between two or more quantities

17.1 What Are Polygons?

Polygons are **plane figures** with three or more sides.

Look at the examples below.

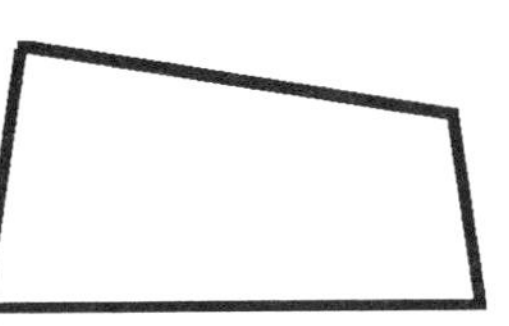

A **quadrilateral** has four sides.

A **triangle** has three sides.

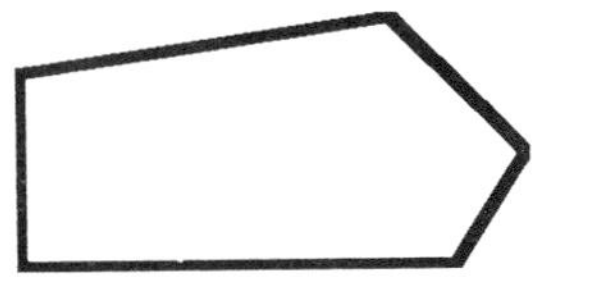

A **pentagon** has five sides.

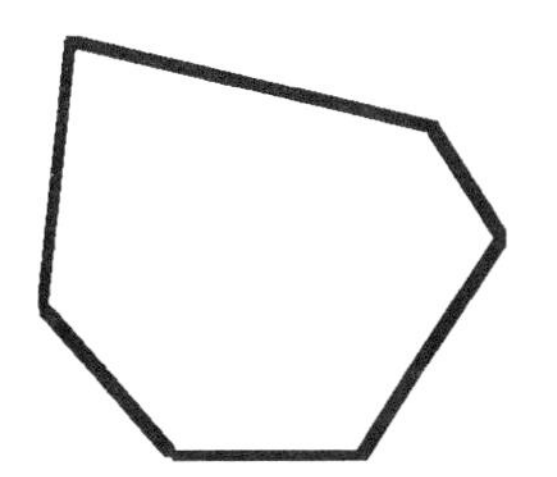

A **hexagon** has six sides.

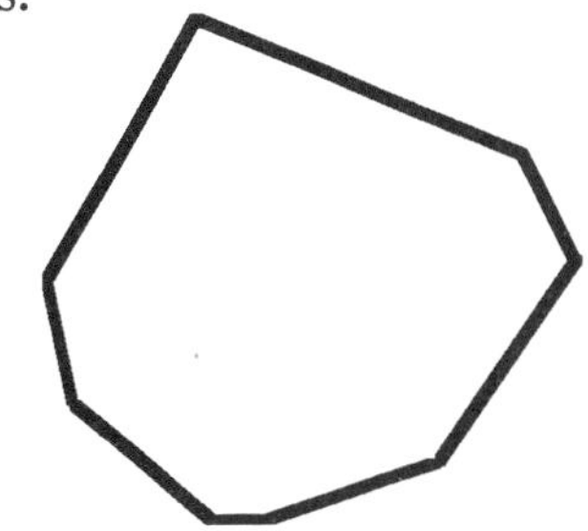

An **octagon** has eight sides.

PRACTICE

Copy the name of each polygon on a sheet of paper. Next to each name write how many sides that polygon has.

1. octagon
2. triangle
3. pentagon
4. quadrilateral
5. hexagon

17.2 Measuring Angles

An **angle** is the space between two straight lines meeting at a point. The size of an angle is measured in **degrees.** The sign for degree or degrees is °. Look at these angles. Each one is a different size.

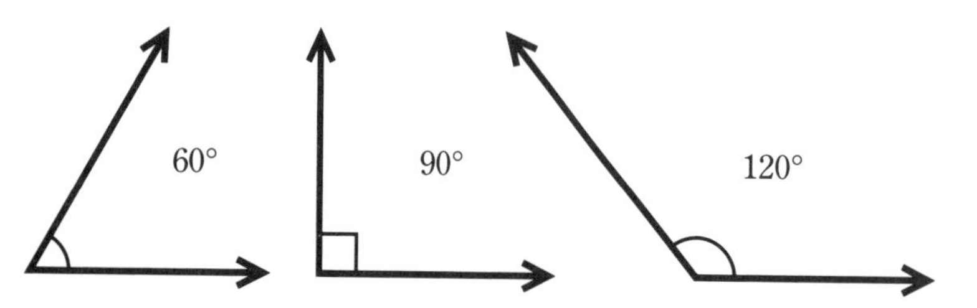

Notice the angle in the center. It measures 90°. It's called a *right angle.* The angle to its left is less than 90°. It's called an *acute angle.* The angle to the right is more than 90°. It's called an *obtuse angle.*

Angles are measured with a protractor. The degree is the unit of measure marked on a protractor. There are 180 degrees on a protractor.

PRACTICE

Copy these angles on a separate sheet of paper. Label each angle acute, right or obtuse.

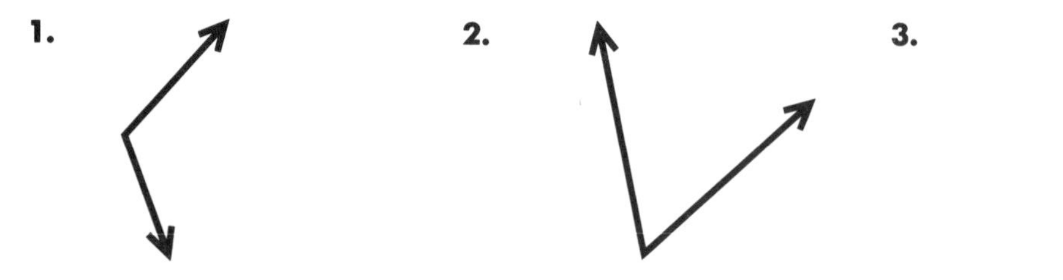

Measuring angles with a protractor is simple. Follow the steps below.

Step 1: Put the center of the protractor on the point where the lines of the angle meet. The bottom of the protractor should be along the bottom line of the angle.

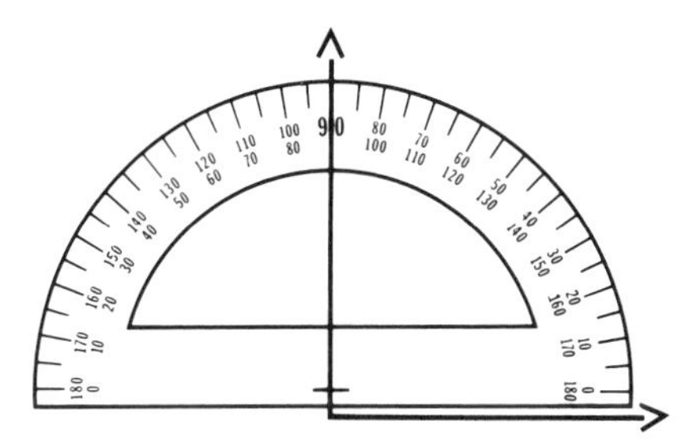

Step 2: A protractor has two scales—an outer scale and an inner scale. To find the number of degrees in the angle, read one of the scales. Use the scale that has a zero where the other line of the angle crosses the scale.

Look at the angles in the margin. The angle above measures 45°. The angle below measures 110°.

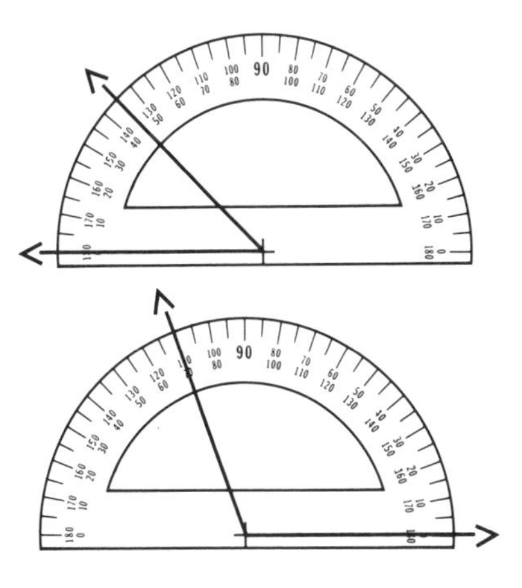

PRACTICE

Use a protractor to measure each angle. Write your answers on a separate sheet of paper.

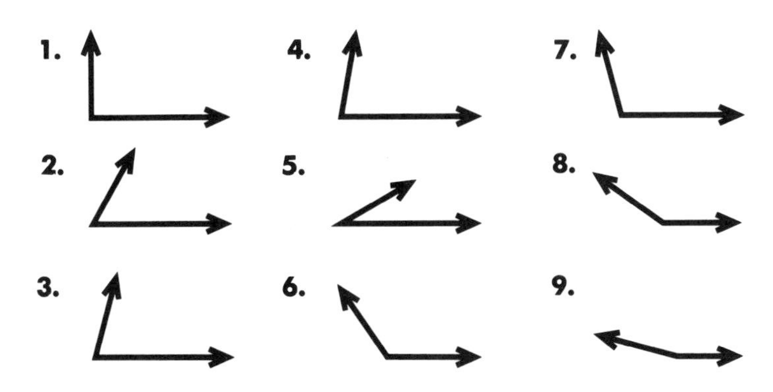

17.3 Quadrilaterals

Quadrilaterals are plane figures with four sides. Study the four different quadrilaterals below.

A **rectangle** has four sides and four right angles. The opposite sides of a rectangle are equal and parallel.

A **square** has four equal sides and four right angles. The opposite sides are parallel.

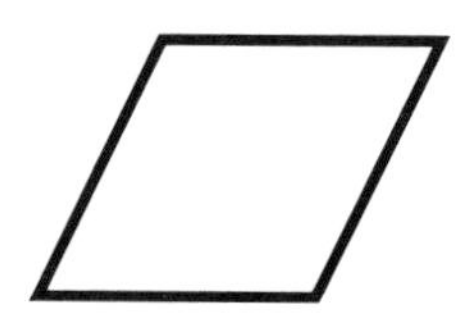

A **parallelogram** has four sides. Its opposite sides are **parallel lines.** Its opposite sides are equal. Its opposite angles are equal. Right angles are not always present.

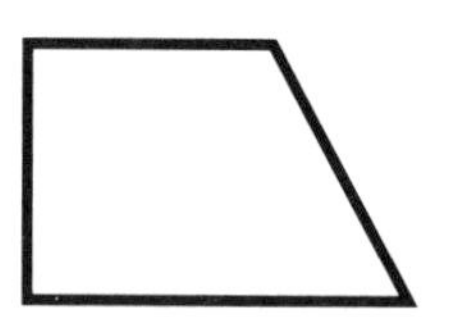

A **trapezoid** has four sides. Two of the sides are parallel.

PRACTICE

Answer the questions below on a separate sheet of paper.

1. Which quadrilaterals have four equal sides?
2. Which quadrilaterals have opposite sides that are equal?
3. Which quadrilaterals have four right angles?

Did You Know?
One of the biggest gems in the world is the Brazilian Princess. It has 221 facets, o sides, and is about as big as a big lemon.

4. Which quadrilaterals have no right angles?
5. Which quadrilaterals have sides that will never meet no matter how far they are extended?
6. Which quadrilateral has a pair of sides that will meet if they are extended far enough?

17.4 Finding the Perimeter

In Chapter 15 you learned about perimeter. You know the perimeter is the boundary or the distance around a plane figure.

So you know you can find the perimeter of any polygon by adding the lengths of each side together.

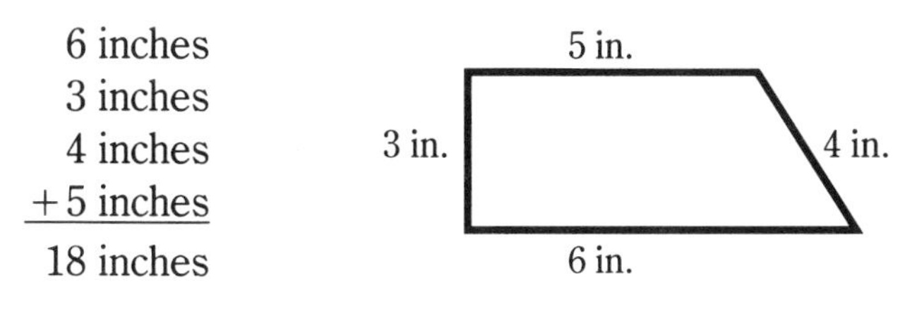

PRACTICE

Find the perimeter of each polygon below. Work on a separate sheet of paper.

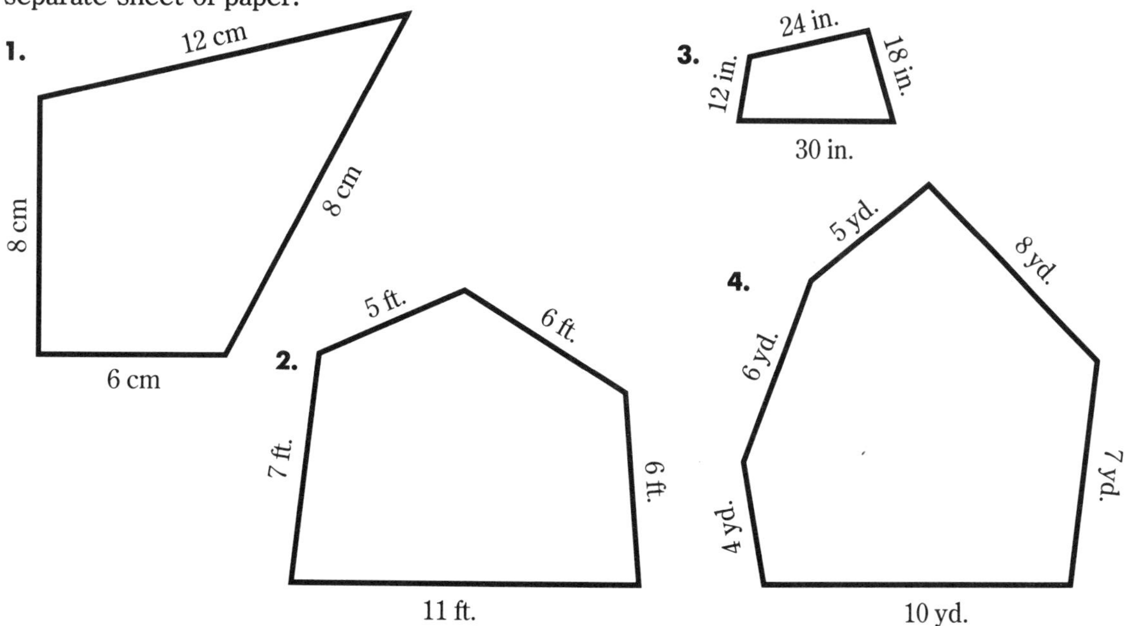

A **formula** is a mathematical rule showing the relationship between two or more quantities. In Chapter 15 you found the perimeter of a rectangle by using this formula:

Perimeter = $(2 \times \text{length}) + (2 \times \text{width})$
$P = (2 \times l) + (2 \times w)$
$P = (2 \times 4) + (2 \times 3)$
$P = 8 + 6$
$P = 14$ inches

4 in.
3 in.
3 in.
4 in.

Squares are special rectangles. In Chapter 15 you found the perimeter of a square by using this formula:

Perimeter = $4 \times \text{side}$
$P = 4 \times s$
$P = 4 \times 3$
$P = 12$

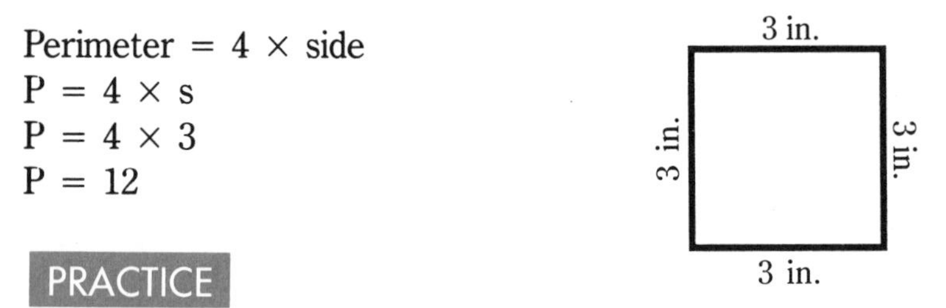

PRACTICE

Find the perimeter of each rectangle or square below. Work on a separate sheet of paper.

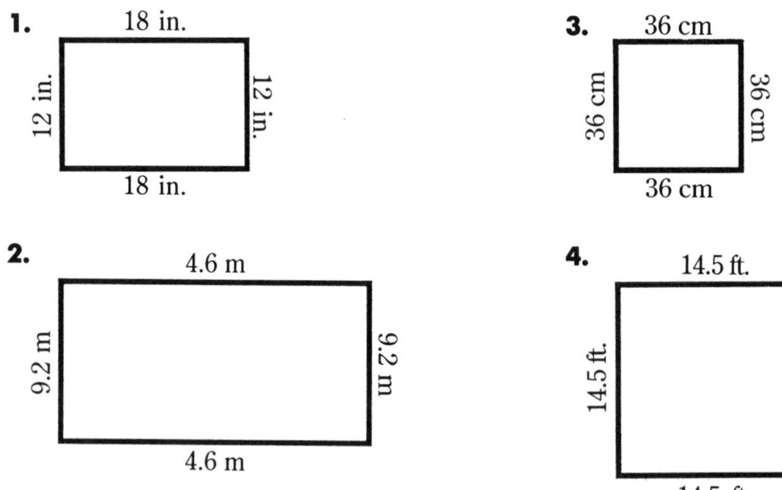

17.5 Areas of Squares and Rectangles

In Chapter 15 you learned that the area of a figure is the amount of surface it covers.

You know that area is measured in square units such as square feet, square miles, square centimeters, and square meters.

The formula we used in Chapter 15 found the area of a rectangle. The formula is:

Area = length × width or $A = l \times w$

Find the area of a rectangle 4 inches by 3 inches.

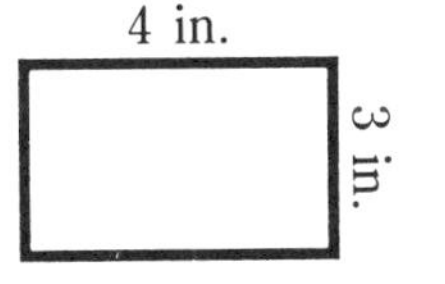

$A = l \times w$
$A = 4 \times 3$
Area = 12 square inches

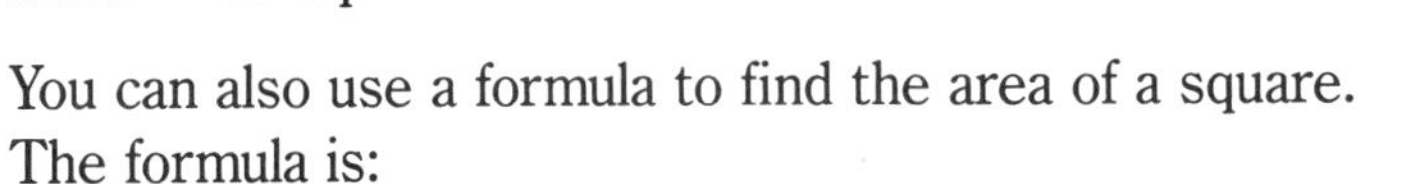

You can also use a formula to find the area of a square. The formula is:

Area = side × side or $A = s \times s$

Find the area of a square 3 feet by 3 feet.

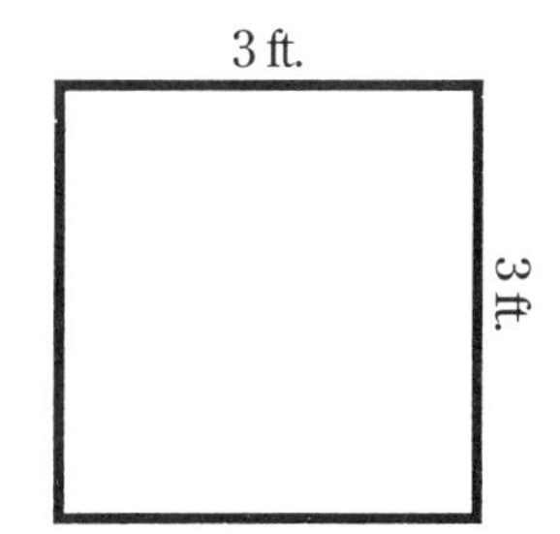

$A = s \times s$
$A = 3 \times 3$
A = 9 square feet

Find the area of a square that is 7 feet by 7 feet.
$A = s \times s$
$A = 7 \times 7$
A = 49 square feet

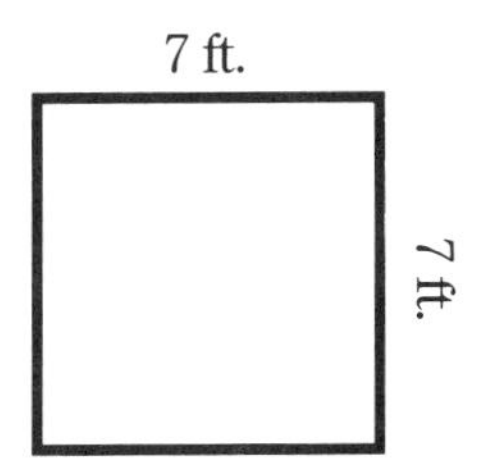

PRACTICE

Find the area of each figure below. Work on a separate sheet of paper.

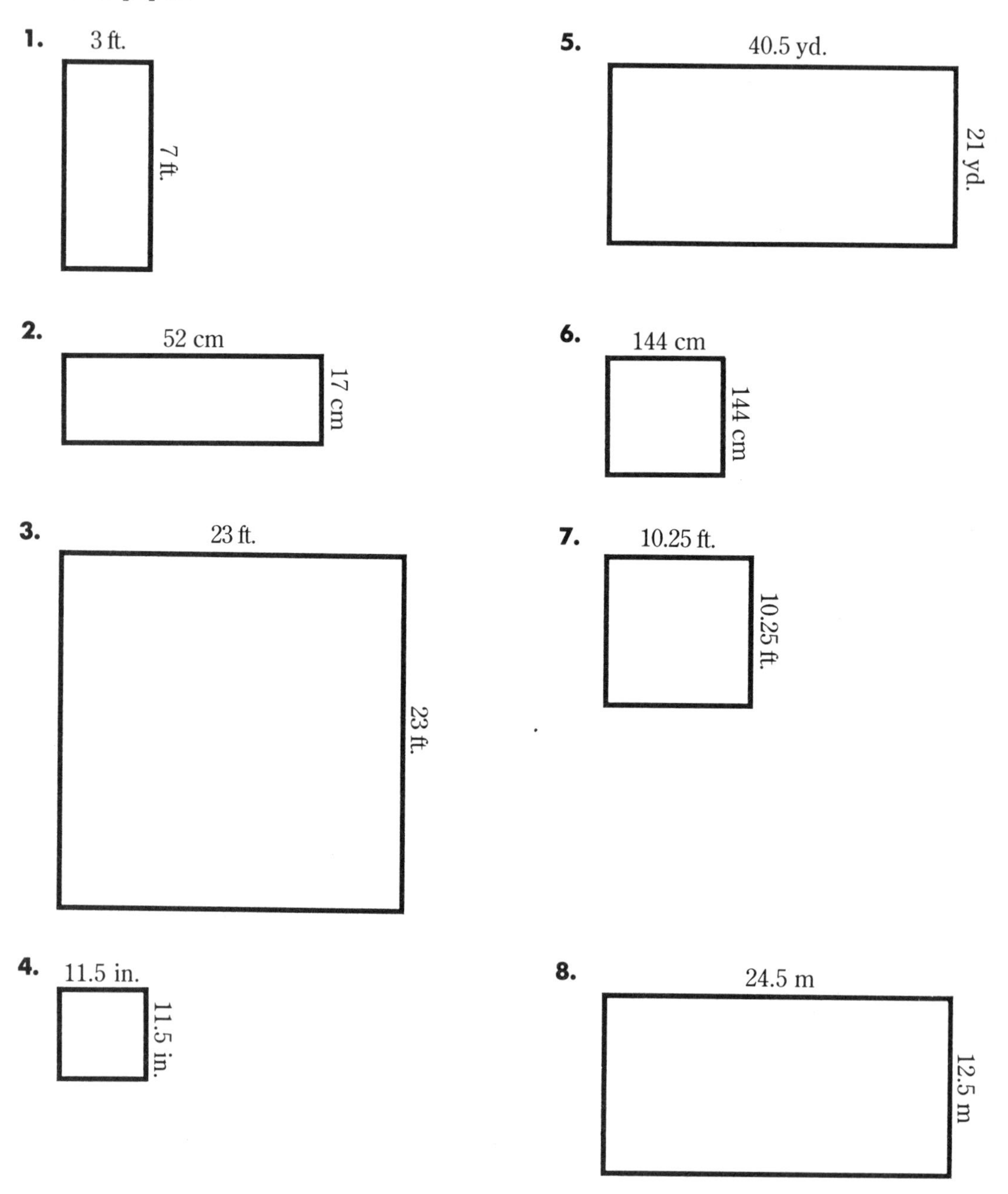

17.6 Area of Parallelograms

Study the picture below. Notice the side labeled base. The height is the distance between the base and the side opposite. Notice where the line to measure the height is drawn. The base must meet the height line in a right (90°) angle.

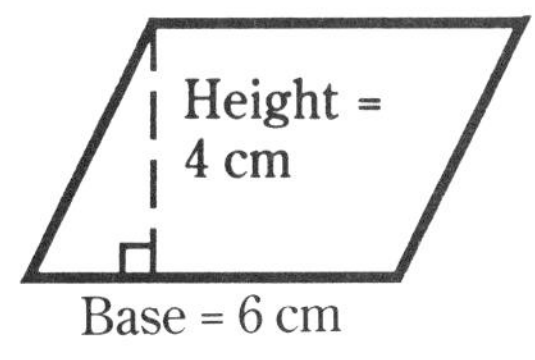

You can find the area of a parallelogram if you know the length of the base and the height.

The formula for the area of a parallelogram is:

$$\begin{aligned} \text{Area} &= \text{base} \times \text{height} \\ A &= b \times h \\ A &= 6 \times 4 \\ A &= 24 \text{ square centimeters} \end{aligned}$$

Remember—before you can find the area, the distances of measurement must be in the same units. Sometimes that means converting one of the measurements before you use the formula.

PRACTICE

A. Find the area of each parallelogram pictured below. Work on a separate sheet of paper.

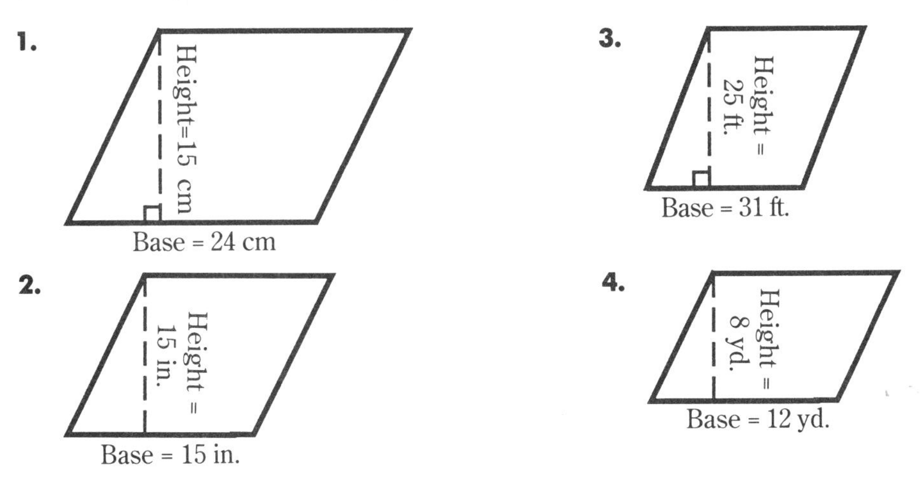

PRACTICE

B. Find the area of each parallelogram described below. Convert the smaller units of measurements to larger units before finding the areas. Work on a separate sheet of paper.

If you need help converting units of measurement, turn back to Chapters 15 and 16.

1. Base = 6 feet
   Height = 36 inches
2. Base = 12 yards
   Height = 24 feet
3. Base = 14 meters
   Height = 800 centimeters
4. Base = 20 yards
   Height = 288 inches
5. Base = 5 dekameters
   Height = 1,000 centimeters
6. Base = 900 centimeters
   Height = 6 meters

17.7 Square Roots

If you know the area of a square, you can find the length of its sides.

16 sq. feet

You know the formula for the area of a square.

$A = s \times s$. Another way to write this is: $A = (s)^2$. This formula reads: Area equals side squared. The 2 means that two s's are multiplied together.

An area that is 16 square feet has sides that are each 4 feet long. Since $4^2 = 16$, 4 is called the square root of 16.

The square root of a number times the square root of a number = the number.

$\sqrt{\ }$ is the sign for square root.

Because $\sqrt{4} \times \sqrt{4} = 4$ and $2 \times 2 = 4$, then $\sqrt{4} = 2$.

Because $\sqrt{9} \times \sqrt{9} = 9$ and $3 \times 3 = 9$, then $\sqrt{9} = 3$.

Because $\sqrt{16} \times \sqrt{16} = 16$ and $4 \times 4 = 16$, then $\sqrt{16} = 4$.

Numbers such as 4, 9, and 16 are called perfect squares because their square roots are whole numbers.

PRACTICE

The square roots of these numbers are whole numbers. Find the square roots by guessing. Then check your answers by multiplying.

1. $\sqrt{25}$ **4.** $\sqrt{49}$ **7.** $\sqrt{225}$

2. $\sqrt{100}$ **5.** $\sqrt{64}$ **8.** $\sqrt{36}$

3. $\sqrt{144}$ **6.** $\sqrt{81}$ **9.** $\sqrt{324}$

Many numbers are not perfect squares. This means that their square roots will not be whole numbers. Notice the square roots below.

$\sqrt{4} = 2$
$\sqrt{5} \cong 2.236$
$\sqrt{6} \cong 2.449$
$\sqrt{7} \cong 2.646$
$\sqrt{8} \cong 2.828$
$\sqrt{9} = 3$

These square roots are between the numbers 2 and 3.

$\cong$ means approximate

You can find the square roots of numbers that are not perfect squares. Check the table on page 350 for the square roots of numbers from 1 to 100.

PRACTICE

Write the square roots of these numbers on a separate sheet of paper. Use the table on page 350.

1. $\sqrt{10}$ **5.** $\sqrt{34}$ **9.** $\sqrt{28}$

2. $\sqrt{13}$ **6.** $\sqrt{21}$ **10.** $\sqrt{83}$

3. $\sqrt{50}$ **7.** $\sqrt{19}$ **11.** $\sqrt{76}$

4. $\sqrt{14}$ **8.** $\sqrt{15}$ **12.** $\sqrt{40}$

17.8 Finding the Angles in a Triangle

A triangle is a plane figure with three sides.

You can remember what an equilateral triangle is by thinking about the word *equal.* An equilateral triangle has three *equal* sides.

There are four kinds of triangles.

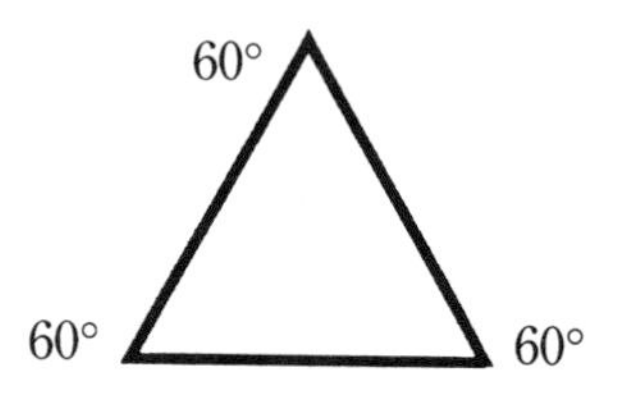

An *equilateral triangle* has three sides of the same length. Each angle measures 60°.

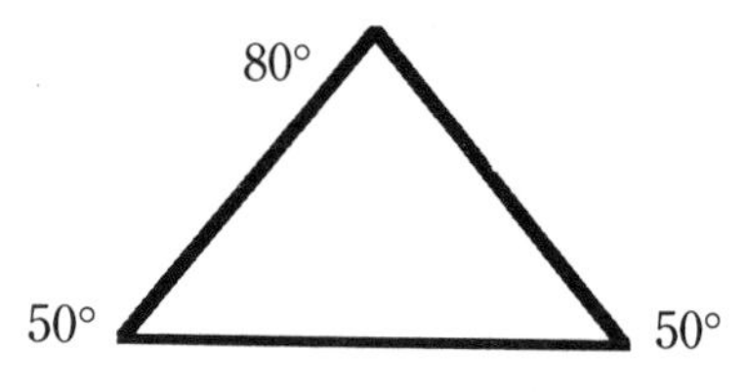

An *isosceles triangle* has two sides of equal length. The angles opposite the equal sides are also equal.

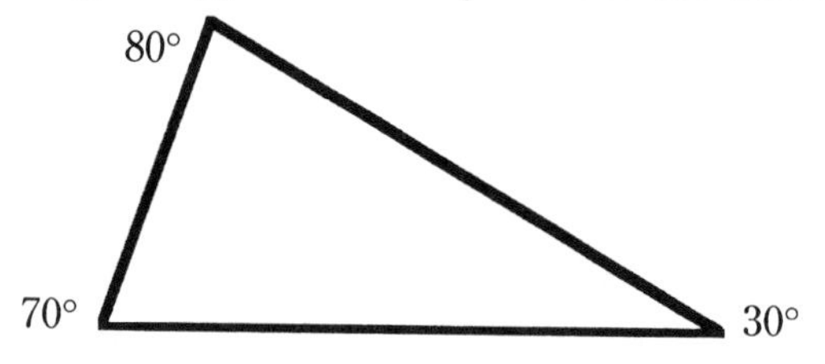

A *scalene triangle* has sides of different lengths. Each angle has a different measure.

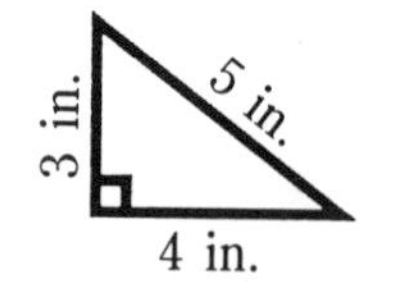

A *right triangle* has one right angle.

The sum of the angles in any triangle is always 180 degrees.

Look at any triangle. If you know the size of two angles you can find the size of the third without measuring.

For example:

Find the size of the third angle in this triangle. Follow the steps below.

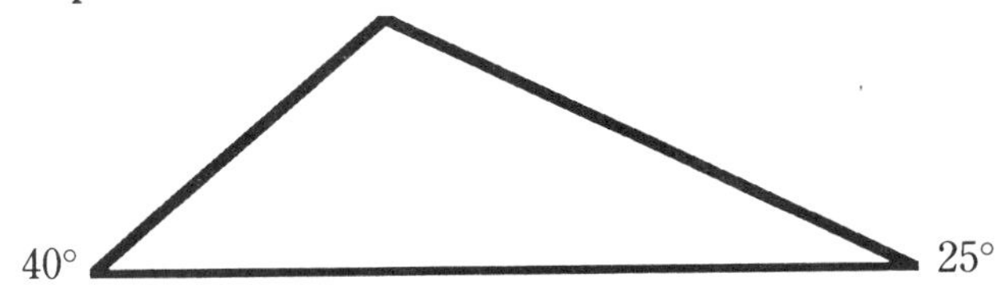

Step 1: Find the sum of the two known angles.

$$\begin{array}{r} 40^\circ \\ +25^\circ \\ \hline 65^\circ \end{array}$$

Step 2: You know that the sum of the three angles in any triangle is 180°. Finding the difference between 180° and 65° will give you the size of the third angle. 115° is the size of the third angle.

$$\begin{array}{r} 180^\circ \\ -\ 65^\circ \\ \hline 115^\circ \end{array}$$

PRACTICE

Find the size of the third angle in each triangle below. Work on a separate sheet of paper.

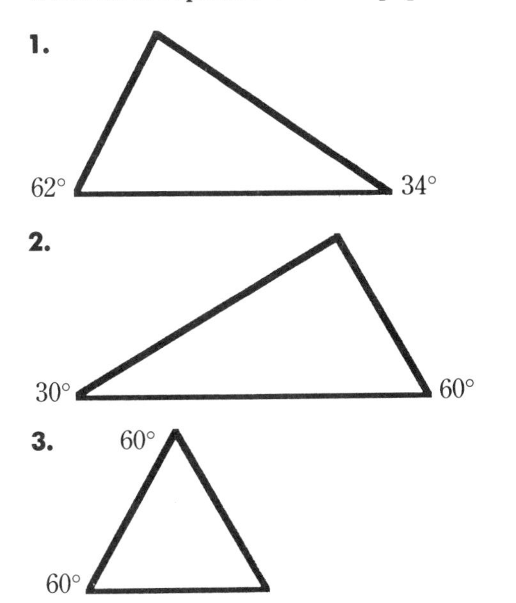

4.

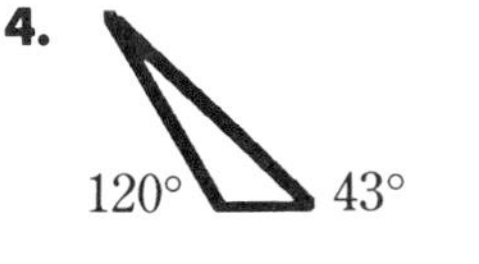

5.

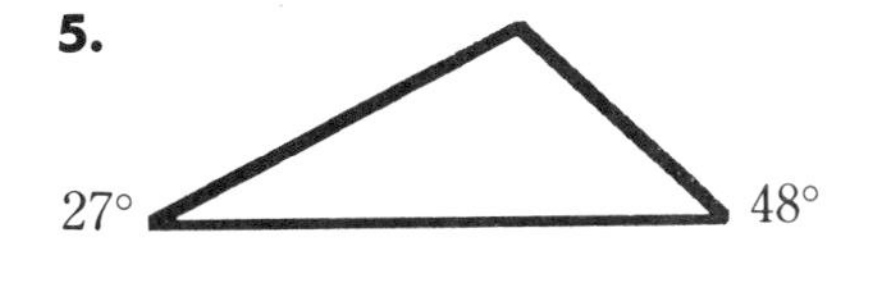

6.

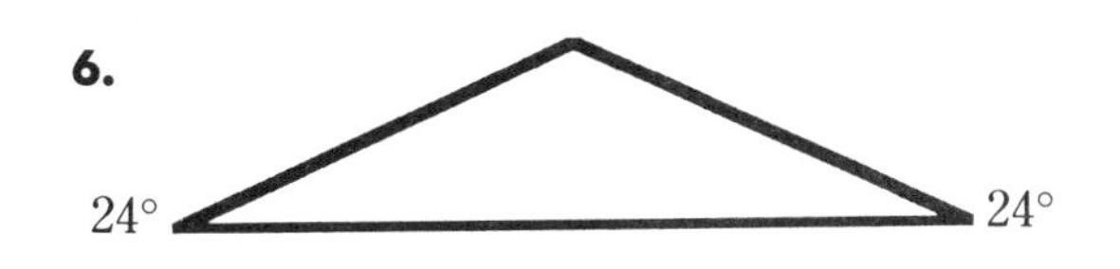

17.9 Finding the Area of a Triangle

You have already learned how to find the area of a parallelogram. The formula is Area = base $\times$ height.

Height = 5 cm

Base = 7 cm

You can also find the area of a triangle.

Think of a triangle as one half of a parallelogram. When you cut a parallelogram in half, you get two triangles.

The formula for finding the area of a triangle is:

Area $= \frac{1}{2} \times$ base $\times$ height

$A = \frac{1}{2} \times b \times h$

$A = \frac{1}{2} \times 7 \times 4$

$A = \frac{1}{2} \times 28$

$A = \frac{28}{2} = 14$

$A = 14$ cm

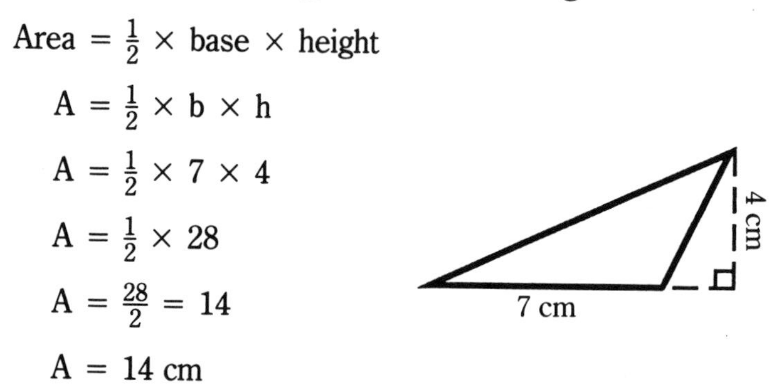

Remember that the height line must meet the base in a right angle (90°). In some triangles this causes the height line to be outside of the triangle.

PRACTICE

A. Find the area of each triangle shown below. Work on a separate sheet of paper.

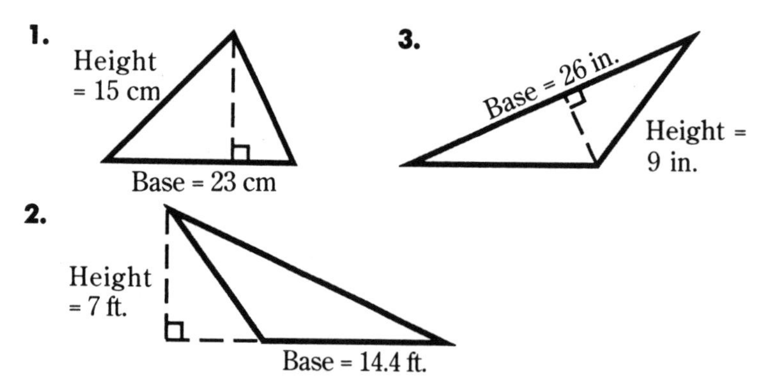

B. Find the area of each triangle described below.

1. Base = 67 centimeters
   Height = 42 centimeters
2. Base = 33 feet
   Height = 20 feet
3. Base = 38.2 centimeters
   Height = 23.6 centimeters
4. Base = 8 yards
   Height = 5 yards
5. Base = 62 inches
   Height = 37 inches
6. Base = 4.5 meters
   Height = 6 meters

17.10 Solving Problems: Perimeter and Area

To avoid confusing perimeter and area, remember to:

1. Read the problem carefully.
2. Decide what you are being asked to find. Is it the outside edge of a figure (perimeter) or the figure's total surface (area)?

PROBLEMS TO SOLVE

Solve each problem below. Work on a separate sheet of paper.

1. A room is 12 feet long and 5 feet wide. What is its area?
2. Mary wants to carpet her living room. The room is 15 feet long and 13 feet wide. How many square feet of carpet will she need?
3. Jerry's yard is 60 feet long and 95 feet wide. How many feet of fencing will he need to enclose the yard?
4. Below is the diagram of a farmer's field. What is the area of the field?

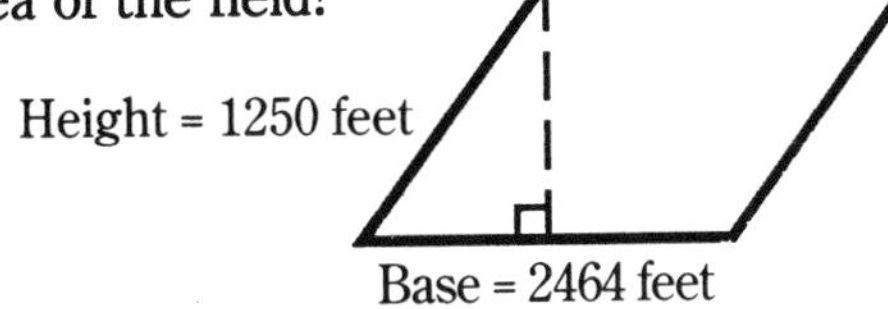

5. Lawrence wants to plant grass in his front yard. One pound of grass seed will cover 500 square feet. His yard is 100 feet long and 150 feet wide. How much grass seed will he need?

MATHEMATICS IN YOUR LIFE:
Home Repair

Ted and Doris decided to lay a new tile floor in their kitchen. Doris measured the floor. She found it was 12 feet by 20 feet.

"The area is 240 square feet. If we get tiles that are one foot square, we'll need 240 tiles," said Ted.

"But the tiles we liked best were only 6 inches square," answered Doris. "How many of those will we need?"

Ted drew this picture.

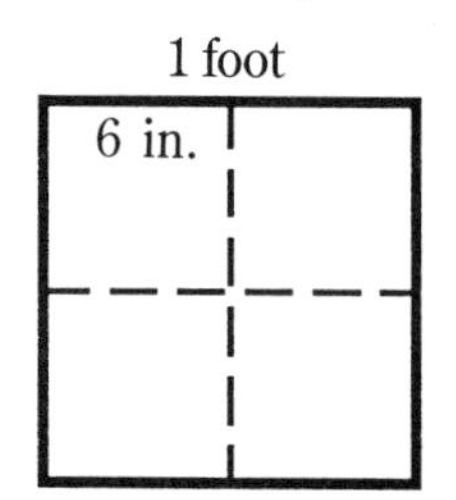

"There are four 6 inch square tiles in 1 square foot," he said.

Doris multiplied quickly. "$4 \times 240 = 960$. We'll need 960 tiles."

1. Suppose Ted and Doris had wanted a tile that was 3 inches square. How many tiles would they have needed?
2. Suppose Ted and Doris decided to use 6-inch tiles in the bathroom. The bathroom floor is 8 feet long and 5 feet wide. How many tiles would they need?
3. Six-inch tiles cost $.29 each. One-foot square tiles cost $.99 each. Which size tiles would cost less to cover Ted and Doris' kitchen floor?

17.11 Using Your Calculator: Perimeter and Area

You can use your calculator to find perimeters and areas.

Solve this problem. Follow the steps below.

A triangle has a base of 78 inches and a height of 69 inches. What is its area?

Step 1: Press [7] [8]

Step 2: Press [×]

Step 3: Press [6] [9] [=]

Step 4: Press [÷]

Step 5: Press [2]

Step 6: Press [=]

Step 7: Your answer should be: 2,691

PRACTICE

Use your calculator to find the area of the triangles described below.

1. A triangle with a base of 46 yards and a height of 53 yards
2. A triangle with a base of 317 feet and a height of 98 feet
3. A triangle with a base of 768 feet and a height of 845 feet

Use your calculator to find the areas of these figures.

4. A rectangle 388 feet by 29 feet.
5. A parallelogram with a base of 96 inches and a height of 67 inches.
6. A square with a side of 256 centimeters.

CHAPTER REVIEW

CHAPTER SUMMARY

- **Polygon** A polygon is a plane figure with three or more sides. Squares, rectangles, parallelograms, and triangles are polygons.
- **Angle** An angle is the space between two straight lines meeting at a point.
- **Degree** A degree is the unit used to measure angles; the symbol for degree is °.
- **Parallel lines** Parallel lines are lines that extend in the same direction and at the same distance apart. Parallel lines can never meet.
- **Parallelogram** A parallelogram is a plane figure with four sides. The opposite sides of a parallelogram are equal.
- **Triangle** A triangle is a plane figure with three sides.
- **Formula** A formula is a mathematical rule showing the relationship between two or more quantities.

REVIEWING VOCABULARY

Number a separate sheet of paper from 1 to 6. Read each word in the column on the left. Find its definition in the column on the right. Write the letter of the definition next to each number.

1. quadrilateral	**a.** a mathematical rule showing the relationship between two or more quantities
2. square	**b.** a plane figure with four sides
3. degree	**c.** a plane figure with three or more sides
4. formula	**d.** a plane figure with four sides and four ninety degree angles
5. polygon	**e.** the unit used to measure angles
6. rectangle	**f.** a plane figure with four equal sides and four ninety degree angles

CHAPTER QUIZ

A. Solve these problems on a separate sheet of paper.

1. Find the perimeter of a square whose side is 12 feet.
2. Find the perimeter of a rectangle whose length is 7 yards and width is 14 yards.
3. Find the area of a square whose side is 26 centimeters.
4. Find the area of a rectangle whose length is 9 meters and width is 13 meters.
5. Find the area of a parallelogram whose base is 48 feet and height is 37 feet.
6. Find the area of a triangle whose base is 63 inches and height is 39 inches.
7. A triangle has angles of 72° and 25°. What is the measure of the third angle?
8. A triangle has angles of 121° and 40°. What is the measure of the third angle?
9. Find the square root of 169 by guessing. Check your answer by multiplying.
10. Find the square root of 484 by guessing. Check your answer by multiplying.

B. Look at the triangles below. On a separate sheet of paper write the name of each triangle and describe its measurements.

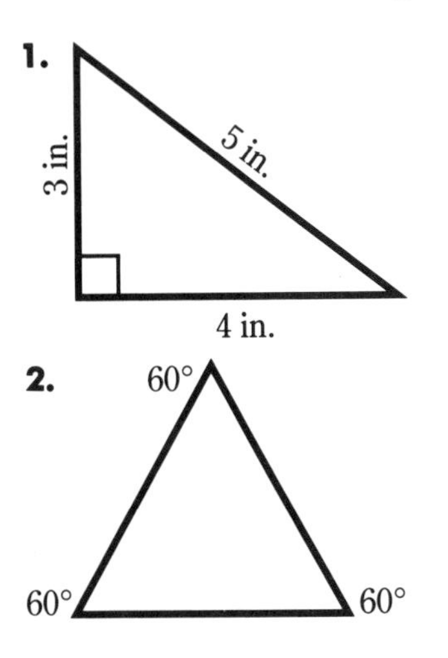

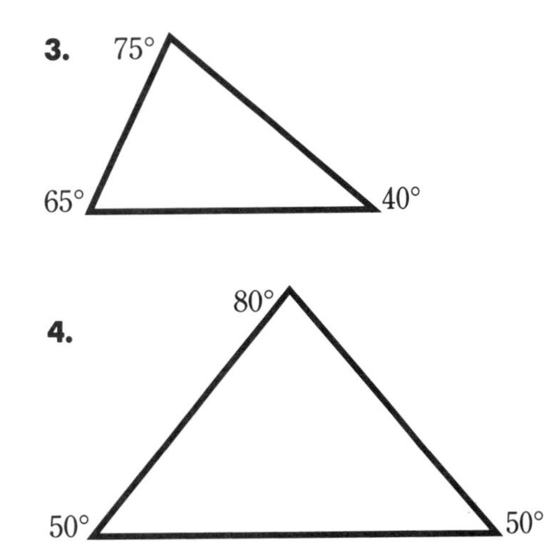

CIRCLES

18

This photograph shows the planet Saturn and its famous ring system. The overall diameter of the rings is 169,000 miles. That's more than twice the diameter of the planet itself.

Chapter Learning Objectives

1. Identify the parts of a circle
2. Explain the meaning of π
3. Find the circumference of a circle
4. Find the area of a circle

WORDS TO KNOW

- **circle** — a closed plane figure; all points on a circle are the same distance from a center point
- **circumference** — the distance or perimeter around a circle
- **diameter** — any straight line across a circle that joins two points of the circle and passes through its center
- **radius** — the distance from the center of a circle to a point on the circle

18.1 The Parts of a Circle

A **circle** is a plane figure with all its points the same distance from a center point.

Study the picture below. Notice the labels.

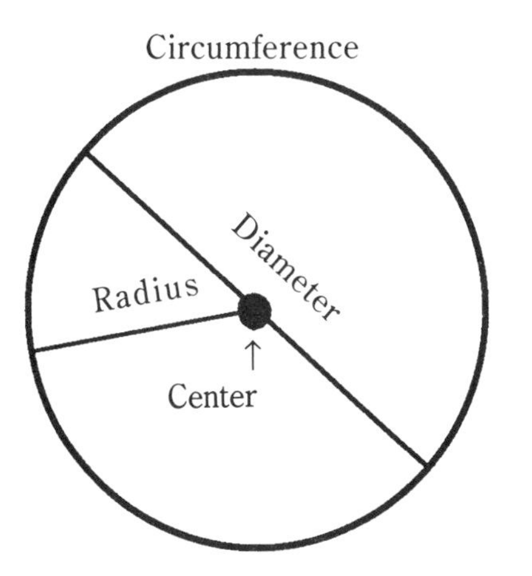

The **circumference** is the distance around a circle.

The **diameter** is any straight line that joins two points of the circle and passes through its center.

The **radius** is the distance from the center of a circle to a point on the circle. The radius of a circle is one-half the length of its diameter.

Did You Know?
The most popular "circular" food in the world is probably pizza. One of the largest pizzas ever baked was *100 feet* in diameter. It was cut into 90,000 slices!

PRACTICE

Number a separate sheet of paper from 1 to 3. Answer these questions about each circle below.

A. What is the diameter?
B. What is the radius?
C. What is the circumference?

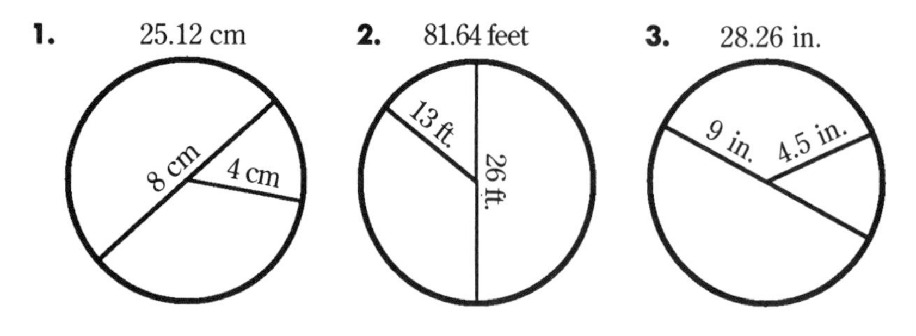

18.2 Circumference of a Circle

A Ratio Called Pi

There is a formula for finding the circumference of a circle. Before you learn it, you need to know about a ratio called *pi*. The Greek letter pi (written π) is the ratio of a circle's circumference to its diameter ($\pi = \frac{c}{d}$). Pi cannot be represented exactly by a decimal or a fraction. Pi is approximately 3.14 or $\frac{22}{7}$.

You can find the circumference of a circle by multiplying its diameter times pi. Find the circumference of this circle. Use the formula below.

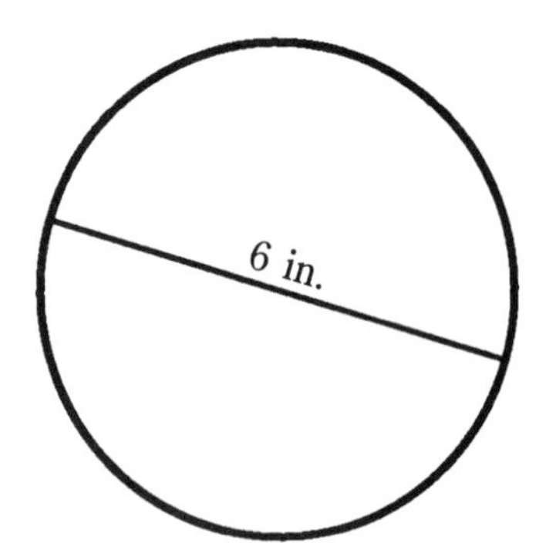

The formula for the circumference of a circle is:

$$\text{Circumference} = \pi \times \text{diameter}$$
$$C = \pi \times d$$
$$C = 3.14 \times 6$$
$$C = 18.84 \text{ in}$$

You have already learned that the radius of a circle is one-half the length of the diameter. This means that the diameter is twice the length of the radius.

You can find the circumference of a circle if you know its radius.

Find the circumference of this circle. Use the formula below.

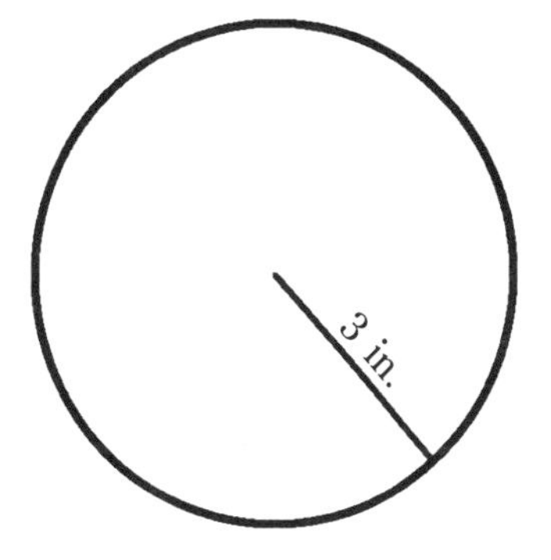

The formula for the circumference of a circle is:

Circumference $= 2 \times \pi \times$ radius

$C = 2 \times \pi \times r$

$C = 2 \times 3.14 \times 3$

$C = 18.84$ in

PRACTICE

Find the circumference of each circle below on a separate sheet of paper.

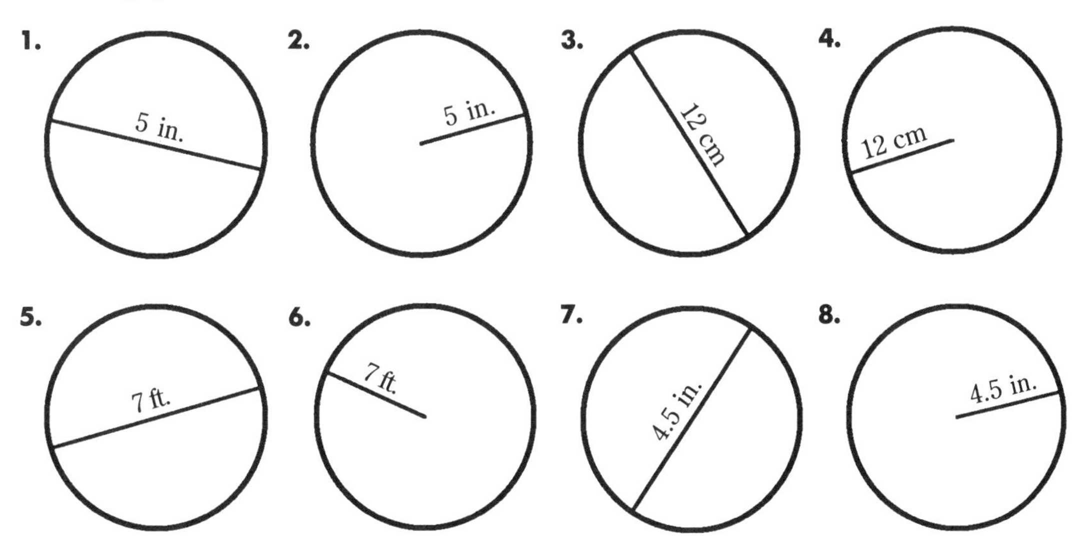

18.3 Finding the Area of a Circle

You can find the area of a circle if you know its radius.

Find the area of this circle. Use the formula below.

The formula for the area of a circle is:

Area = $\pi \times$ radius squared

$A = \pi \times r^2$

$A = 3.14 \times 4 \times 4$

$A = 3.14 \times 16$

$A = 50.24$ square inches

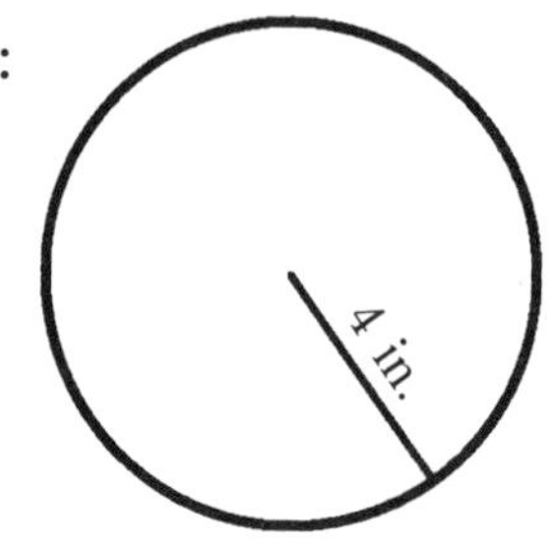

PRACTICE

A. Find the area of a circle for each radius given below.
Work on a separate sheet of paper.

1. radius: 6 cm

2. radius: 9 in

3. radius: 6.5 ft

4. radius: 83 yd

5. radius: 27 in

6. radius: 90 cm

B. Copy the chart below on a separate sheet of paper.
Fill in the missing information.

Radius	Diameter	Circumference	Area
	12 in		
37 cm			
2.7 cm			
	9 ft		
	49 in		
$6\frac{1}{2}$ ft			
		37.68 m	
			50.24 sq. in

18.4 Solving Problems: Circles

To find the circumference of a circle, you need to know either its radius or its diameter. The two formulas for finding circumference are:

$C = \pi \times D$
$C = 2 \times \pi \times R$

To find the area of a circle, you need to know its radius or diameter.

If you know the radius, use the formula $A = \pi r^2$.

If you know the diameter, you must first find the radius.

You can do this by dividing the diameter by 2. Then use the formula $\pi \times r^2$.

PROBLEMS TO SOLVE

Solve the problems below on a separate sheet of paper.

1. A radio station broadcasts over a radius of 94 miles. What is the size of the area that receives its programs?
2. A table has a diameter of 5 feet. How large is its area?
3. A person can see 4 miles in any direction from the tower on top of Mount Gray. How large is the area that can be seen?
4. Manny wants to make a round tabletop. The distance from the center of the wood to the nearest edge is 2 feet. How big a tabletop can he make?
5. A round table has a diameter of 13 feet. If 20 people are seated around it, how much room does each person have at the edge of the table?
6. A sprinkler sprays water 18 feet in every direction. How much ground does the sprinkler cover?
7. Hanna wants to plant tulips around the edge of her round flower bed. The flower bed has a diameter of 14 feet. She will plant the tulips six inches apart. How many tulips will she need?

MATHEMATICS IN YOUR LIFE: Making Tablecloths

Elizabeth makes circular tablecloths. She always wants to make the largest cloth she can from a piece of fabric.

Here is how she gets ready to make a tablecloth.

First, she measures to find the center of the fabric. This length gives her the radius of the largest circle she can cut. She then knows how large the tablecloth will be.

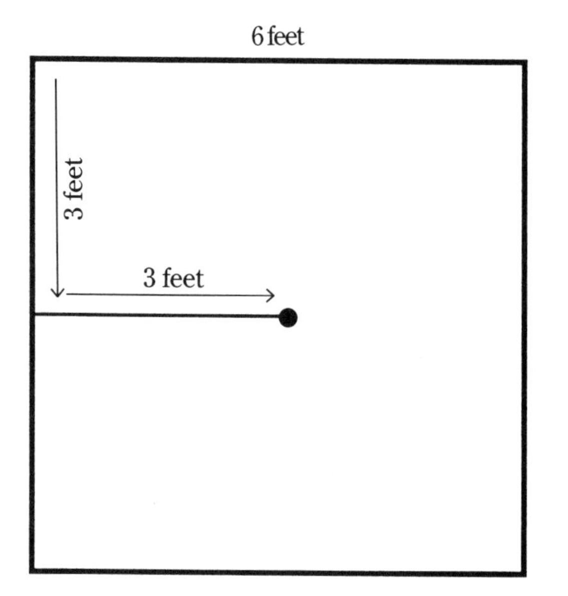

1. What is the largest tablecloth that Elizabeth can make from the material shown above?
2. How much fringe will Elizabeth need to sew around the circumference of the tablecloth?
3. Suppose that Elizabeth had fabric that measured 9 feet from the center to the edge. How big a tablecloth could she make from it?
4. How much fringe will Elizabeth need to sew around the circumference of the tablecloth in question 3?

18.5 Using Your Calculator: Finding the Diameter

You've already learned how to find the circumference of a circle. You need to know either the radius or diameter and use the correct formula with π.

You can also find the diameter of a circle when you know its circumference. The formula for diameter is:

$$D = \frac{C}{\pi}$$

You can use your calculator to help you find the diameter quickly. For example: Find the diameter of a circle with a circumference of 25.12 inches.

Step 1: Press: [2] [5] [.] [1] [2]

Step 2: Press: [÷]

Step 3: Press: [3] [.] [1] [4]

Step 4: Press: [=]

Step 5: Your answer should be: 8

PRACTICE

Use your calculator to find the diameters of circles with the following circumferences. Take your answers to the hundredth decimal place.

1. 20.41 feet

2. 57 centimeters

3. 65.25 yards

4. 99 inches

5. 47.10 feet

6. 75.65 centimeters

7. 172.7 inches

8. 188.4 yards

9. 81.64 meters

10. 314 millimeters

11. 125.6 kilometers

12. 106.76 feet

CHAPTER REVIEW

CHAPTER SUMMARY

- **Circle** — A circle is a closed plane figure. All points on a circle are the same distance from a center point.
- **Circumference** — The circumference of a circle is the distance around the circle.
- **Diameter** — The diameter of a circle is a straight line. It passes through the center of the circle, connecting two points on the circumference.
- **Radius** — The radius of a circle is a straight line. It connects the center with a point on the circumference.
- **π (Pi)** — Pi is a ratio. It cannot be represented exactly by a fraction or a decimal. Pi is approximately 3.14 or $\frac{22}{7}$. Pi is part of the formula for finding the circumference of a circle. Pi is also part of the formula for finding the area of a circle.

REVIEWING VOCABULARY

Copy the sentences below. Put the correct word from the box in each blank.

circumference	π (Pi)	diameter	radius

1. The ________ is a straight line that connects the center of a circle with a point on the circle.
2. The ________ is the distance around a circle.
3. The ________ passes through the center of a circle, connecting two points of the circle.
4. ________ is part of the formula for finding the area and circumference of a circle.

CHAPTER REVIEW

CHAPTER QUIZ

A. Find the circumferences and areas of the circles shown below.

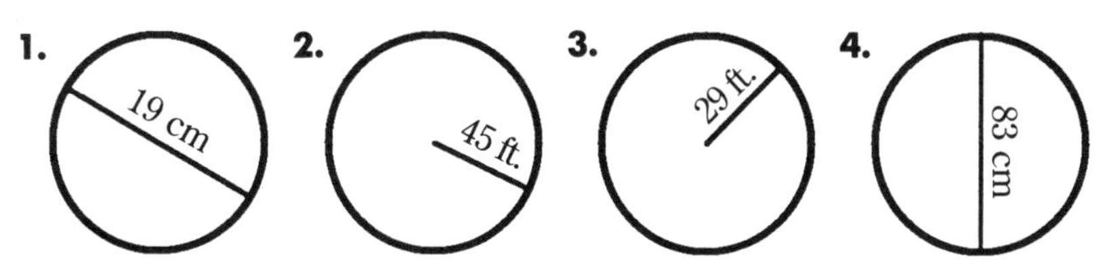

B. Developing Thinking Skills
Read each problem and study the pictures below. Then answer the questions on a separate sheet of paper.

1. A clock's radius is 5 inches. How much area does the minute hand sweep over in 30 minutes?

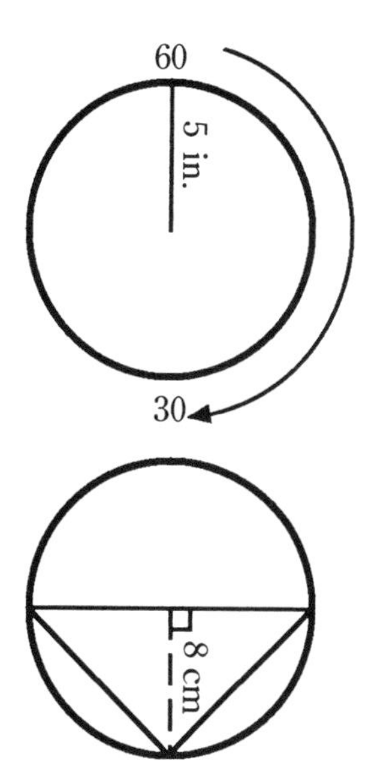

2. The radius of the circle is 8 centimeters. What is the area of the triangle?

C. Extra Challenge
Read the problems below carefully. Solve them on a separate sheet of paper.

1. A circular table is 3 feet in diameter and 3 feet in height. How many square feet of cloth is needed to cut a tablecloth for the entire table. Let the tablecloth touch the floor. (Hint: First you need to find the radius of the table.)
2. A circular walkway will enclose a circular playground. The walkway is 3 feet wide. The diameter of the playground is 40 feet. What is the area of the walkway? (Hint: You'll need to find the area of the playground first.)

EQUATIONS

19

For nearly 15 years, Boston Red Sox outfielder Jim Rice has been one of baseball's best hitters. Baseball uses various methods to determine how well players perform. In this chapter one of the things you'll learn is how to figure out a player's batting average.

Chapter Learning Objectives

1. Identify true and false equations
2. Explain the meaning of parentheses in equations
3. Perform operations in the correct order
4. Solve equations with one operation
5. Solve equations with more than one operation
6. Identify exponents
7. Solve word problems using equations

WORDS TO KNOW

- **equation** — a mathematical sentence stating that two quantities are equal
- **variable** — a letter that stands for an unknown number
- **open sentence** — an equation that contains at least one variable
- **expression** — one or more numerals or variables joined by operational symbols $(+, -, \times, \div)$

19.1 What Is an Equation?

An **equation** is a mathematical sentence stating that two quantities are equal. An equation can be true or false.

true equation: $2 + 3 = 5$
false equation: $7 - 1 = 8$

Some equations contain **variables.** A variable is a letter that stands for a number.

Equations that contain variables are called **open sentences.**

open sentence: $x + 3 = 9$
$8 - x = 6$

variable: $x + 3 = 9$ variable: $8 - x = 6$

A variable can be replaced by a number. Then the equation will be either true or false. So open equations may be true for some number replacements and false for others. Study these equations.

open sentence: $x + 3 = 9$
true equation: $6 + 3 = 9$
false equation: $5 + 3 = 9$

The open sentence becomes true when x is replaced by 6.

The open sentence becomes false when x is replaced by 5.

PRACTICE

On a separate sheet of paper rewrite each equation twice.

First replace the variable by a number that makes the equation true. Then replace the variable by a number that makes the equation false.

1. $x + 4 = 13$
2. $9 + x = 17$
3. $27 - x = 3$
4. $19 - x = 13$
5. $25 + x = 48$
6. $53 - x = 41$
7. $3 \times x = 21$
8. $x \times 8 = 40$
9. $89 - x = 53$

19.2 Parentheses in Equations

Parentheses in equations can mean several things.

Parentheses around an operation means "Do this operation first." Study the examples below.

	$(7 + 4) - 6 =$
Add first.	$11 - 6 =$
Next subtract.	$11 - 6 = 5$
	$5 + (2 \times 6) =$
Multiply first.	$5 + 12 =$
Next add.	$5 + 12 = 17$

PRACTICE

Solve these equations on a separate sheet of paper.

1. $(3 + 9) - 7 = b$
2. $38 - (9 \times 2) = s$
3. $(7 \times 8) - 26 = w$
4. $93 - (63 - 9) = n$
5. $(5 + 7) - (6 + 3) = p$
6. $(8 - 5) \times 6 = a$
7. $(9 \times 8) - 25 = m$
8. $(24 + 8) \times 9 = y$
9. $27 + (5 \times 12) = x$
10. $(8 \times 4) - (3 \times 7) = w$

Parentheses can also tell you to multiply.

$2(6)$ means 2×6

When an equation has a variable, the parentheses are sometimes left out.

$3(x) = 3x = 3 \times x$
$x(y) = xy = x \times y$

Study these examples.

Here the parentheses tell you to multiply.

$4(8) = 32$

Here the parentheses tell you to add before multiplying.

	$3(6 + 1) =$
Add first.	$3(7) =$
Next multiply.	$3(7) = 21$

PRACTICE

Solve these equations on a separate sheet of paper.

1. $3(8) = x$
2. $59(2) = y$
3. $4(15 - 8) = w$
4. $28 - (7 + 5) = s$
5. $12(8) - 4(7) = p$
6. $(9 - 4) = m$
7. $2(9 - 7) = f$
8. $7(16 + 4) = p$
9. $2(12) - 13 = d$
10. $5(9) + 3(3) = n$

19.3 Order of Operations

Some **expressions** do not contain parentheses that tell you what to do first. Then you must do the operations in this order:

An expression, like an equation, is a problem to be solved.

1. Multiply and divide in order from left to right.
2. Add and subtract in order from left to right.

Study these examples. Do the operations in order.

Step 1: Check to see what operations must be done. In this expression there is no multiplication or division. You need to add and subtract from left to right.

$3 + 6 - 4 =$

Step 2: Add 3 + 6.

$3 + 6 = 9$

Step 3: Subtract 4 from 9.

$9 - 4 = 5$

$3 + 6 - 4 = 5$

Step 1: Check to see what operations must be done. In this expression you must divide *before* you can subtract from left to right.

$16 - 10 \div 2 =$

Step 2: Divide 10 by 2.

$10 \div 2 = 5$

Step 3: Subtract 5 from 16.

$16 - 5 = 11$
$16 - 10 \div 2 = 11$

PRACTICE

Solve the equations below on a separate sheet of paper. Remember the order of operations.

1. $7 + 6 - 2 = m$
2. $29 - 3 \times 5 = p$
3. $8 + 19 - 1 = s$
4. $52 - 18 \div 6 = w$
5. $17 \times 8 \times 9 = c$
6. $42 \times 6 + 6 = a$
7. $37 - 7 \times 3 = s$
8. $33 \div 11 \times 6 = b$
9. $64 \div 2 + 5 = n$
10. $54 + 64 \div 8 = d$

19.4 Solving Equations with Addition and Subtraction

You can solve an equation by finding all of the numbers that make the equation true.

You can solve equations with addition and subtraction. Solve $x + 8 = 15$. Follow the steps below.

Step 1: Think about what the equation means. This equation means that 8 plus an unknown number equals 15.

$x + 8 = 15$

Step 2: Undo the result of adding 8 by subtracting 8 from both sides. This gives you the unknown number.

$$\begin{array}{rcr} x + 8 & = & 15 \\ -8 & & -8 \\ \hline x & = & 7 \end{array}$$

(Remember: Any operation done to one side of the equation must also be done to the other.)

Step 3: Check your answer. Replace the variable in the equation with the number 7 to make the equation true.

$x + 8 = 15$
$7 + 8 = 15$

Solve $y - 3 = 4$. Follow the steps below.

Step 1: Think about what the equation means. This equation means that 3 subtracted from an unknown number equals 4.

$$y - 3 = 4$$

Step 2: Undo the result of subtracting 3 by adding 3 to both sides. This gives you the unknown number.

(Remember: Any operation done to one side of the equation must also be done to the other.)

$$\begin{array}{rcr} y - 3 & = & 4 \\ +3 & & +3 \\ \hline y & = & 7 \end{array}$$

Step 3: Check your answer. Replace the variable in the equation with the number 7 to make the equation true.

$$y - 3 = 4$$
$$7 - 3 = 4$$

PRACTICE

Solve these equations on a separate sheet of paper. Check your work.

1. $9 + c = 29$
2. $n - 5 = 34$
3. $x + 32 = 33$
4. $x + 45 = 62$
5. $92 + m = 105$
6. $n - 25 = 90$
7. $45 - x = 37$
8. $21 - a = 16$
9. $81 + x = 210$

19.5 Solving Equations with Multiplication and Division

You solve an equation by finding all of the numbers that make the equation true.

You can solve equations with multiplication and division. Solve $24 = 4x$. Follow the steps below.

Multiplying or dividing by zero results in zero. That's why you never multiply or divide either side of an equation by zero.

Step 1: Think about what the equation means. This equation means that 4 multiplied by an unknown number equals 24.

$$24 = 4x$$

Step 2: Undo the result of multiplying by 4 by dividing both sides by 4. This gives you the unknown number.

(Remember: Any operation done to one side of the equation must also be done to the other.)

$$\frac{24}{4} = \frac{4x}{4}$$
$$6 = x$$

Step 3: Check your answer. Replace the variable in the equation with the number 6 to make the equation true.

$24 = 4x$
$24 = 4 \times 6$

Solve $\frac{y}{5} = 4$. Follow the steps below.

Step 1: Think about what the equation means. This equation means that an unknown number divided by 5 equals 4.

$\frac{y}{5} = 4$

Step 2: Undo the result of dividing by 5 by multiplying both sides by 5. This gives you the unknown number.

(Remember: Any operation done to one side of the equation must also be done to the other.)

$5\left(\frac{y}{5}\right) = 5(4)$
$y = 20$

Step 3: Check your answer. Replace the variable in the equation by the number 20 to make the equation true.

$\frac{y}{5} = 4$
$\frac{20}{5} = 4$

PRACTICE

Solve these equations on a separate sheet of paper. Check your answers.

1. $4n = 16$

2. $30 = 6c$

3. $\frac{c}{4} = 14$

4. $\frac{n}{5} = 9$

5. $18 = \frac{x}{8}$

6. $9y = 114$

7. $15b = 120$

8. $65 = \frac{x}{3}$

9. $162 = 3x$

19.6 Solving Equations with More Than One Operation

Sometimes you need to do more than one operation to solve an equation.

You know that you do operations in a certain order. When you undo operations, you undo them in the reverse order.

Doing Operations	Undoing Operations
1. Multiply and divide in order from left to right. **2.** Add and subtract in order from left to right.	**1.** Add and subtract in order from left to right. **2.** Multiply and divide in order from left to right.

Solve this equation. Follow the steps below.

Step 1: Think about what the equation means. This equation means that 3 subtracted from an unknown number multiplied by 4 equals 5.

$$4x - 3 = 5$$

Step 2: Undo the result of subtracting 3 by adding 3 to both sides.

$$\begin{aligned} 4x - 3 &= 5 \\ +3 \quad & \quad +3 \\ 4x &= 8 \end{aligned}$$

Step 3: Undo the result of multiplying by 4 by dividing both sides by 4. This gives you the unknown number.

$$\frac{4x}{4} = \frac{8}{4}$$
$$x = 2$$

Step 4: Check your answer. Replace the variable in the equation by the number 2 to make the equation true.

$$4x - 3 = 5$$
$$4 \times 2 - 3 = 5$$
$$8 - 3 = 5$$

PRACTICE

Solve these equations on a separate sheet of paper. Remember the order of undoing operations. Check your answers.

1. $4a + 7 = 19$

2. $23 - 5n = 13$

3. $2(13) - y = 20$

4. $8 + 22 = 6x$

5. $24 = \frac{x}{3} + 6$

6. $3 + 9 = \frac{x}{3}$

7. $\frac{n}{2} + 6 = 28$

8. $48 = 7y + 13$

9. $6m - 4 = 14$

10. $5k + 3(3) = 29$

19.7 Exponents

Some expressions ask you to multiply a number by itself one or more times. Study the chart below.

This	means	and	is read
2^1	2	two	to the first power
2^2	2×2	two	squared
2^3	$2 \times 2 \times 2$	two	cubed
2^4	$2 \times 2 \times 2 \times 2$	two	to the fourth power
2^5	$2 \times 2 \times 2 \times 2 \times 2$	two	to the fifth power
2^6	$2 \times 2 \times 2 \times 2 \times 2 \times 2$	two	to the sixth power

The smaller, raised number is called an *exponent*. Using exponents is a useful way to write large numbers.

$3^3 = 3 \times 3 \times 3 = 27$

$5^4 = 5 \times 5 \times 5 \times 5 = 625$

$6^5 = 6 \times 6 \times 6 \times 6 \times 6 = 7776$

You can solve problems with exponents.

Solve $4^3 - 3^2 + 3 = a$ Follow the steps below.

Step 1: Multiply the numbers with exponents.

Step 2: Do the other operations in order.

$4^3 - 3^2 + 3 =$
$4 \times 4 \times 4 = 64$
$3 \times 3 = 9$
$64 - 9 =$
$64 - 9 = 55$
$55 + 3 = 58$
$a = 58$
$4^3 - 3^2 + 3 = 58$

Here is another example.

Solve $36 - 4^2 + 8 - 2^2 = x$

Step 1: Multiply the numbers with exponents.

$36 - 4^2 + 8 - 2^2 =$
$4 \times 4 = 16$
$2 \times 2 = 4$

Step 2: Do the other operations in order.

$36 - 16 + 8 - 4 =$
$20 + 8 - 4 =$
$28 - 4 = 24$
$x = 24$

PRACTICE

Solve these equations on a separate sheet of paper.

1. $5^3 + 6^2 - 2 = y$

2. $8^2 - 3^4 = a$

3. $4^5 - 5^4 = c$

4. $6^3 \times 2^4 + 8 = w$

5. $3^5 \times 4^4 = s$

6. $8^2 \times 5^3 = x$

7. $45^2 \div 2 = p$

8. $5^2 \times 3^5 \div 6 = c$

19.8 Solving Problems: Equations

You can make up equations to solve problems.

Problem: Pedro and Martin have 36 dollars between them.

Pedro has 3 times as many dollars as Martin has. How many dollars does each man have?

Making up the equation: Let d stand for the number of dollars Martin has.
Let 3d stand for the number of dollars Pedro has.

$$d + 3d = 36$$
$$4d = 36$$
$$\frac{4d}{4} = \frac{36}{4}$$
$$d = 9 \text{ (the number of dollars Martin has)}$$
$$3d = 27 \text{ (the number of dollars Pedro has)}$$

Make up an equation and solve this word problem.

Susan and Ellen have worked 54 hours between them. Susan has worked twice as many hours as Ellen has. How many hours has each woman worked?

MATHEMATICS IN YOUR LIFE: Batting Averages

Batting averages are used to compare how well different players hit. They are also used to compare a player's current performance with his or her past performances.

To find a batting average, you need to know two things.

—how many times the player has been at bat
—how many hits he or she has made

The formula is $\frac{\text{hits}}{\text{times at bat}}$ = batting average or $\frac{h}{t} = b$.

Batting averages are expressed in decimals and carried to 3 places.

A player has come to bat 30 times. He has made 15 hits. What is his batting average?

$\frac{15}{30} = b$

$30\left(\frac{15}{30}\right) = 30(b)$

$15 = 30b$

$\frac{1}{2} = b$

$.500 = b$

$\frac{15}{30} = .500$

1. Jed Davis made 204 hits in 498 times at bat. What is his batting average?
2. Carla Trantior came to bat 286 times. Her batting average for the season was .391. How many hits did she get?
3. Jennifer made 32 hits in 85 times at bat. What was her batting average?
4. Tom came to bat 300 times. His batting average was .400. How many hits did he get?

19.9 Using Your Calculator: Solving Equations

You can use your calculator to help you solve equations. You just have to enter the operations in the right order.

You can press [×] [÷] [+] and [−] until all the operations are done. Then press [=] .

Solve this equation: $5x - 3 = 27$

Step 1: Press [2] [7]

Step 2: Press [+]

Step 3: Press [3]

Step 4: Press [÷]

Step 5: Press [5]

Step 6: Press [=]

$x = 6$

Check your answer

Step 1: Press [5]

Step 2: Press [×]

Step 3: Press [6]

Step 4: Press [−]

Step 5: Press [3]

Step 6: Press [=] 27 is the value of the original equation

Step 7: Your answer should be: $x = 6$

Use your calculator to solve these problems. Check your answers.

1. $7y + 4 = 46$

2. $3(x) - 2 = 22$

3. $6 + 8n = 78$

4. $9 + 53 = 6y$

5. $15t - 7 = 83$

6. $3(s) - 24 = 90$

CHAPTER REVIEW

CHAPTER SUMMARY

■ Equation	An equation is a mathematical sentence stating that two quantities are equal. An equation may be true or false.
■ Variable	A variable is a letter that stands for an unknown number.
■ Open sentence	An open sentence contains at least one variable. An open sentence is neither true nor false. Replacing variables by a number makes an open equation true or false.
■ Parentheses	Parentheses around an operation tell you to do this operation first. Parentheses around a number or variable tell you to multiply.
■ Order of Operations	Operations must be done in this order. **1.** Multiply and divide in order from left to right. **2.** Add and subtract in order from left to right.
■ Exponent	An exponent tells how many times a number is multiplied by itself.

REVIEWING VOCABULARY

Copy the sentences on a separate sheet of paper. Put the correct word from the box in each blank.

variable	equation	open sentence	expression

1. An ________ contains a variable.

2. A ________ is a letter that stands for an unknown number.

3. A mathematical sentence that states that two quantities are equal is an ________ .

4. One or more numerals or variables joined by operational symbols is an ________ .

Chapter Review

CHAPTER QUIZ

A. Solve these equations on a separate sheet of paper. Check your answers.

1. $49 - (8 \times 3) = y$
2. $(6 + 8) - (5 - 2) = s$
3. $2(5 + 4) = w$
4. $n - 7 = 24$
5. $120 = 35 + x$
6. $23 + (32 \div 4) = d$
7. $(15 - 9) + (6 \times 7) = a$
8. $6(8 + 5) = y$
9. $5n = 35$
10. $7b - 12 = 37$
11. $5a + 9 = 34$
12. $9z = 72$
13. $6^5 - 5^5 = s$
14. $\frac{y}{27} = 108$
15. $y + 5 = 57$
16. $7^5 \times 4^4 = s$

B. Developing Thinking Skills

Solve the following word problems by writing equations for them. Do your work on a separate sheet of paper.

1. Jim is four times older than Steve. The difference in their ages is 24 years. How old is Steve? How old is Jim?
2. Mrs. Hunt paid $639 for her daughter's discount airline ticket. The daughter's fare was $\frac{3}{4}$ of the adult fare. How much did Mrs. Hunt pay for her own fare? How much was the cost of both tickets together?
3. Jeff has three times as many library books as Jerry. The sum of their books is 48. How many books does Jeff have? How many books does Jerry have?
4. Melissa paid $16 for a cassette storage case that was on sale for 20% off. What was the original price of the cassette case? (Change the percent to a decimal.)

UNIT SIX REVIEW

Solve the problems below on a separate sheet of paper.

1. Find the perimeter of a square whose side is 67 centimeters.
2. Find the perimeter of a rectangle whose length is 9 feet and width is 5 feet.
3. Find the area of a square whose side is 26 yards.
4. Find the area of a rectangle whose length is 5 miles and width is 8 miles.
5. Find the area of a parallelogram whose base is 59 meters and whose height is 42 meters.
6. Find the area of a triangle whose base is 36 inches and whose height is 47 inches.
7. A triangle has angles of 62° and 31°. What is the measure of the third angle?
8. Find the square root of 324.
9. The diameter of a circle is 259 centimeters. What is the radius?
10. The diameter of a circle is 17 feet. What is the circumference?
11. The radius of a circle is 38 inches. What is the circumference?
12. The radius of one circle is 6 feet. The radius of a second circle is 4 feet. Which circle has the longer circumference? How much longer is that circumference?

Solve the following equations on a separate sheet of paper.

13. $39 + (28 \div 7) = y$
14. $(49 - 8) + (9 \times 8) = d$
15. $7(9 + 6) = w$
16. $z + 29 = 73$
17. $58 = 95 - n$
18. $8z = 96$
19. $4c - 13 = 39$
20. $6^4 + 3^3 = p$

GLOSSARY

add to put numbers together; to find the total amount or sum

angle the area between two rays meeting at a point

area the surface a shape covers

arithmetic operation addition, subtraction, multiplication, or division; each one is a separate arithmetic operation

bar graph a graph that uses bars of different lengths to compare amounts or sizes of things

Celsius a scale of measuring temperature

centi- a prefix that stands for .01 (one-hundredth); a *centigram* is one-hundredth of a gram

circle a closed plane figure with all points the same distance from a center point

circle graph a circle-shaped graph that shows how a total amount has been divided into parts

circumference the distance or perimeter around a circle

column numbers placed one below the other

compare to see how two things are alike or different

composite number a number with more than two factors

convert to change into a different form; to exchange for something of equivalent value

cross cancel to divide a numerator of one fraction and a denominator of another fraction by the same number

cross multiply to multiply each of the numerators of a pair of fractions by the denominator of the other

cross product the result of cross multiplying

customary something that is done or based on accepted practice; usual or commonly used

deci- a prefix that stands for .1 (one-tenth); a *decimeter* is one-tenth of a meter

decimal a number written with a dot followed by places to the right; the digits to the right of the dot stand for less than a whole

decimal point a period placed to the left of a decimal

degree the unit used to measure angles; also the unit used to measure temperature; the symbol for degree is °

deka- a prefix that stands for 10; a *dekaliter* is 10 liters

denominator the bottom number in a fraction; the denominator tells how many equal parts there are in the whole unit

diameter any straight line across a circle that joins two points of the circle and passes through its center

difference the amount by which one number is larger or smaller than another; the amount that remains after one number is subtracted from another

digit the symbols used to write numbers: 0, 1, 2, 3, 4, 5, 6, 7, 8, and 9

divide to find out how many times one number contains another

dividend the number to be divided in a division problem

divisible can be divided evenly without a remainder

divisor the number to divide by in a division problem

equal the same as: =

equation a mathematical sentence stating that two quantities are equal

equivalent fractions fractions whose numbers are different but whose values are the same; $\frac{2}{8}$ and $\frac{1}{4}$ are equivalent fractions

estimate to quickly figure out an answer that is close to the exact answer; to make a good guess

even number a number that ends in 0, 2, 4, 6, or 8

expression one or more numerals or variables joined by operational symbols (+, −, ×, ÷)

factor one of the numbers that must be multiplied to obtain a product

Fahrenheit a scale of measuring temperature

formula a mathematical rule showing the relationship between two or more quantities

fraction a number that expresses part of a whole unit; $\frac{2}{3}$ is a fraction

frequency the number of times a number or a range of numbers occurs in a set

gram a metric unit of measurement used to measure weight

graph a picture that shows information as number facts

greatest common factor the largest factor that two or more products share; remember that a factor is one of the numbers that must be multiplied to obtain a product

hecto- a prefix that stands for 100; a *hectogram* is 100 grams

infinite without end

invert to reverse the positions of the numerator and denominator of a fraction

kilo- a prefix that stands for 1,000; a *kilometer* is 1,000 meters

least common multiple the smallest multiple that two or more numbers share

like fractions fractions with the same denominator

like mixed numbers mixed numbers whose fractions have the same denominator

line graph a graph that uses line segments to show changes and relationships between things usually over a period of time

liter a metric unit of measurement used to measure capacity

lowest terms a fraction is in its lowest terms when only 1 divides evenly into the numerator and denominator

mean the number obtained by dividing the sum of a set of quantities by the number of quantities in the set; also called the *average*

median the middle number of a set of numbers when the numbers are arranged in order of size

meter a metric unit of measurement used to measure length

metric system a system of measurement based on the number 10

milli- a prefix that stands for .001 (one-thousandth); a *milliliter* is one-thousandth of a liter

mixed decimal a number containing a whole number and a decimal number

mixed number a number made up of a whole number and a fraction

mode the number in a set of numbers that occurs most frequently

multiples of a number the products of multiplying that number by whole numbers

multiplication a quick way to add

multiply to add a number to itself one or more times

number line numbers in order shown as points on a line

numerator the top number in a fraction; the numerator tells how many parts of the whole unit are being used

odd number a number that ends in 1, 3, 5, 7, or 9

open sentence an equation that contains at least one variable

parallel lines lines that extend in the same direction and at the same distance apart; parallel lines can never meet

parallelogram a plane figure with four sides; the opposite sides of a parallelogram are parallel

partial product the number obtained by multiplying a number by only one digit of a number

percent a part of a whole that has been divided into 100 equal parts

perimeter the distance or boundary around a shape

plane figure a figure that has only length and width

polygon a plane figure with three or more sides

prime number a number with only two factors—itself and 1

probability the likelihood that an event will occur

product the number obtained by multiplying

proportion a statement that two ratios are equal

proportion sign (::) a double colon used between two ratios to show that they are equal

quadrilateral a plane figure with four sides

quotient the number obtained by dividing one number into another; the answer in a division problem

radius the distance from the center of a circle to a point on the circle

range the difference between the highest and lowest numbers in a set

rate a measured amount in relation to a fixed quantity of something else; a *rate of speed* is how fast something is going

ratio a comparison of one number to another

ratio sign (:) a colon used between two numbers to show a ratio

rectangle a plane figure with four sides and four ninety degree angles; the opposite sides of a rectangle are equal

remainder the number left over in a division problem

rename to break down a number into its parts; 28 can be renamed as 2 tens and 8 ones

result the answer

round to change a number to the nearest ten, hundred, thousand, and so on

solve to find the answer to a problem

square a plane figure with four equal sides and four ninety degree angles

statistics numerical information

subtract to take away one number from another; to find the difference between two numbers

sum the amount obtained by adding; the total

term any of the numbers in a ratio or proportion

triangle a plane figure with three sides

unlike fractions fractions with different denominators

unlike mixed numbers mixed numbers whose fractions have different denominators

variable a letter that stands for a number

volume the space or capacity inside a container

whole number 0, 1, 2, 3, 4, 5, 6, 7, and so on

Chapter One Additional Practice

Write the numbers in each group from largest to smallest. Work on a separate sheet of paper.

1. 8 6 3 2 9 4
2. 45 53 72 49 71 69

Write only the odd numbers.

3. 3 5 4 36 59 88 21 72 67 54 91

Copy the sentences below on a separate sheet of paper. Put the correct number words in the blanks.

4. 6,827 means ____ thousands, ____ hundreds, ____ tens, + ____ ones.
5. 7,049 means ____ thousands, ____ hundreds, ____ tens, + ____ ones.

Read the place name before each number. Copy the number on a separate sheet of paper. Then circle the digit that appears in that place.

6. Ones 4,687
7. Hundreds 3,459
8. Thousands 8,318
9. Tens 5,702

On a separate sheet of paper, write the smallest number in each group.

10. 692 340 27 5,350
11. 6,000 2,761 899 600
12. 9,948 17,121 28,560 40,010

On a separate sheet of paper, write the largest number in each group.

13. 769 417 99 2,032
14. 79 692 700 587
15. 76,835 49,092 15,005 75,835

Round each number below to the nearest hundred. Write your answers on a separate sheet of paper.

16. 457
17. 849
18. 2,802
19. 843
20. 7,066
21. 777
22. 998
23. 6,251
24. 1,500

Chapter Two Additional Practice

Copy the problems below on a separate sheet of paper. Find the sums.

1.
```
  3
  9
  7
 +2
```

2.
```
  8
  5
  7
 +9
```

3.
```
  6,529
    481
  7,358
 +8,901
```

4.
```
  3,029
  2,706
  5,035
 +  518
```

5.
```
  9,395
  1,147
  8,716
 +1,012
```

6.
```
  4,110
    931
  5,874
 +9,996
```

Coleen orders the office supplies for her department each month. First, she reads the Supply Request forms that each person fills out. Next she adds the items to find how many of each item she needs to order. Then she enters the numbers on a Department Supply Request form.

Add the requests for each item. Then copy the form below and fill in the totals.
Mark wanted: 4 legal pads, 6 pencils, 3 pens, 1 package of typing paper.
Anne wanted: 2 legal pads, 2 pencils, 4 pens, 3 packages of typing paper.
Marlene wanted: 1 steno pad, 5 pencils, 2 pens, 5 packages of typing paper.
Norm wanted: 6 legal pads, 7 pencils, 1 package of typing paper.

Department Supply Request

Legal pads	
Steno pads	
Packages of typing paper	
Pens	
Pencils	

Chapter Three Additional Practice

Copy the problems below on a separate sheet of paper. Find the differences.

1. $\begin{array}{r} 9{,}827 \\ -3{,}215 \\ \hline \end{array}$
2. $\begin{array}{r} 8{,}392 \\ -2{,}670 \\ \hline \end{array}$
3. $\begin{array}{r} 7{,}843 \\ -6{,}539 \\ \hline \end{array}$
4. $\begin{array}{r} 8{,}756 \\ -\ \ 444 \\ \hline \end{array}$
5. $\begin{array}{r} 7{,}382 \\ -\ \ 754 \\ \hline \end{array}$
6. $\begin{array}{r} 8{,}272 \\ -7{,}358 \\ \hline \end{array}$
7. $\begin{array}{r} 9{,}548 \\ -6{,}305 \\ \hline \end{array}$
8. $\begin{array}{r} 9{,}137 \\ -\ \ 642 \\ \hline \end{array}$
9. $\begin{array}{r} 4{,}560 \\ -1{,}292 \\ \hline \end{array}$
10. $\begin{array}{r} 8{,}657 \\ -4{,}606 \\ \hline \end{array}$
11. $\begin{array}{r} 9{,}753 \\ -2{,}485 \\ \hline \end{array}$
12. $\begin{array}{r} 4{,}931 \\ -\ \ 972 \\ \hline \end{array}$

Solve the following problems on a separate sheet of paper.

David works in the shipping room of a warehouse. At the end of each day, he must write down how many of the following items are left. At the beginning of the week, the warehouse held: 5,829 cartons of typing paper, 6,024 cartons of envelopes, and 7,963 cartons of writing paper.

1. On Monday 492 cartons of typing paper, 1,017 cartons of envelopes, and 976 cartons of writing paper were shipped. How many cartons of each item were left?
2. On Tuesday 905 cartons of typing paper, 917 cartons of envelopes, and 842 cartons of writing paper were shipped. How many cartons of each item were left?
3. On Wednesday 631 cartons of typing paper, 2,010 cartons of envelopes, and 342 cartons of writing paper were shipped. How many cartons of each item were left?
4. On Thursday 503 cartons of typing paper, 782 cartons of envelopes, and 521 cartons of writing paper were shipped. How many cartons of each item were left?
5. On Friday 388 cartons of typing paper, 177 cartons of envelopes, and 539 cartons of writing paper were shipped. How many cartons of each item were left at the end of the week?

Chapter Four Additional Practice

Copy the problems below on a separate sheet of paper. Find the products.

1. 62×4
2. 732×32
3. 70×21
4. 832×30
5. 800×57
6. 516×65
7. 853×84
8. 719×452
9. $7{,}829 \times 2{,}354$
10. $6{,}305 \times 4{,}829$
11. $1{,}392 \times 6{,}502$
12. $4{,}890 \times 3{,}700$

Solve the following problems on a separate sheet of paper.

Lisa was checking supplies as they were delivered to her company's supply room.

1. Four cartons of index cards were delivered. There were 16 boxes in each carton. How many boxes of index cards were delivered?
2. There were already 49 boxes of index cards in the supply room. After the new delivery, how many were there altogether?
3. Eight cartons of typing paper were delivered. There were 12 packages in each carton. How many packages of typing paper were delivered?
4. There were already 67 packages of typing paper in the supply room. After the new delivery, how many packages were there altogether?
5. Later Mark took 54 packages to the word processing room. How many packages were left in the supply room?
6. Lisa unpacked 5 boxes of typewriter ribbons. There were 32 ribbons in each box. How many ribbons were there in all?

Estimate the answers to these problems. First round each number. Then multiply to find the product.

1. 38×53
2. 46×64
3. 753×25
4. 696×509

Chapter Five
Additional Practice

Copy the problems below on a separate sheet of paper. Find the quotients.

1. $4\overline{)488}$

2. $2\overline{)806}$

3. $6\overline{)326}$

4. $8\overline{)4,808}$

5. $7\overline{)4,837}$

6. $27\overline{)324}$

7. $100\overline{)8,921}$

8. $100\overline{)7,500}$

9. $35\overline{)9,931}$

10. $52\overline{)4,836}$

11. $29\overline{)4,289}$

12. $359\overline{)49,785}$

Solve the problems below on a separate sheet of paper.

1. Seth has 444 boxes of greeting cards to pack in cartons. Each carton holds 12 boxes. How many cartons will he need?

2. Jed and Dan loaded 2,240 pounds of cement onto their truck. Each bag weighs 56 pounds. How many bags did they load onto the truck?

3. There are 326 children in Tyler Nursery School. At the beginning of the year, each child is given a box of crayons. Sixteen boxes of crayons come in one carton. How many cartons does the school need?

4. The children have lunch at the Tyler School. There are places for eight children at each table. How many tables does the school need?

5. The school uses 5 cartons of drawing paper a month. There are 12 packages in a carton. How many packages does the school need for 12 months? About how many packages does each child use in a year?

6. Susan and Barney spent 5 days on vacation. They drove 1,255 miles altogether. How many miles did they average each day?

7. Sidney collects stamps. He has a stamp album with 800 pages in it. He has a total of 46,000 stamps. How many stamps are on one page? Estimate your answer.

Chapter Six
Additional Practice

Write the Greatest Common Factor of each pair of numbers below on a separate sheet of paper.

1. 84, 92 **2.** 32, 88 **3.** 54, 71 **4.** 9, 33

Write the Least Common Multiple of each pair of numbers below on a separate sheet of paper.

5. 5, 65 **6.** 21, 87 **7.** 6, 24 **8.** 50, 25

Find the prime factors of each number below. Use factor trees. Work on a separate piece of paper.

9. 28 **11.** 48 **13.** 72 **15.** 68 **17.** 30

10. 21 **12.** 5 **14.** 32 **16.** 31 **18.** 25

Copy the chart below on a separate sheet of paper. Use divisibility tests to decide whether the numbers are divisible by 3,4,6, and 9. Then write either Yes or No in the appropriate column.

Number	Divisible by 3	Divisible by 4	Divisible by 6	Divisible by 9
576				
927				
6,436				
8,112				
17,333				
24,428				

Chapter Seven Additional Practice

Copy the numbers of the pictures below on a separate sheet of paper. Next to each number write the fraction that is represented by the colored part of each shape.

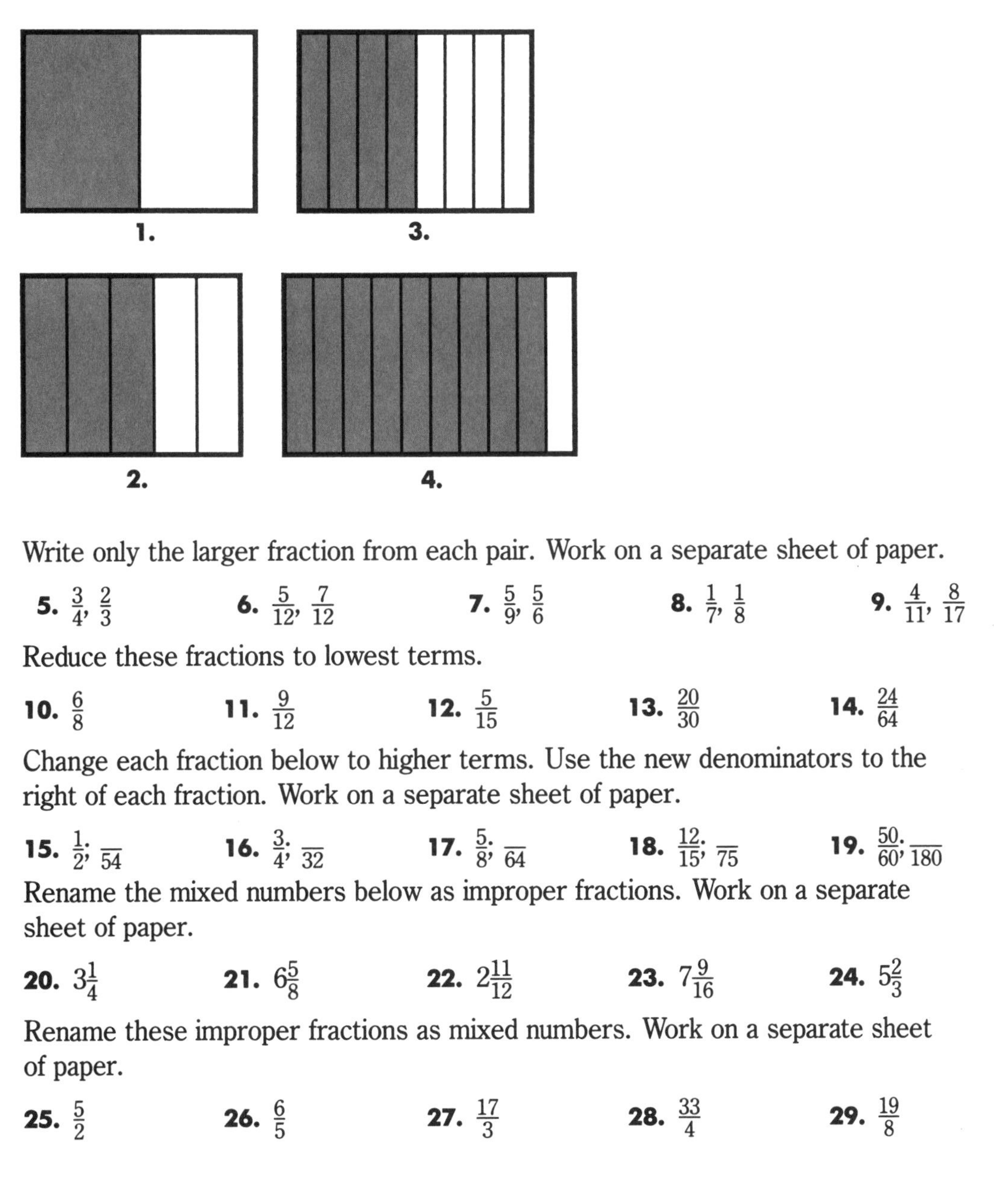

Write only the larger fraction from each pair. Work on a separate sheet of paper.

5. $\frac{3}{4}, \frac{2}{3}$ **6.** $\frac{5}{12}, \frac{7}{12}$ **7.** $\frac{5}{9}, \frac{5}{6}$ **8.** $\frac{1}{7}, \frac{1}{8}$ **9.** $\frac{4}{11}, \frac{8}{17}$

Reduce these fractions to lowest terms.

10. $\frac{6}{8}$ **11.** $\frac{9}{12}$ **12.** $\frac{5}{15}$ **13.** $\frac{20}{30}$ **14.** $\frac{24}{64}$

Change each fraction below to higher terms. Use the new denominators to the right of each fraction. Work on a separate sheet of paper.

15. $\frac{1}{2}, \frac{\quad}{54}$ **16.** $\frac{3}{4}, \frac{\quad}{32}$ **17.** $\frac{5}{8}, \frac{\quad}{64}$ **18.** $\frac{12}{15}, \frac{\quad}{75}$ **19.** $\frac{50}{60}, \frac{\quad}{180}$

Rename the mixed numbers below as improper fractions. Work on a separate sheet of paper.

20. $3\frac{1}{4}$ **21.** $6\frac{5}{8}$ **22.** $2\frac{11}{12}$ **23.** $7\frac{9}{16}$ **24.** $5\frac{2}{3}$

Rename these improper fractions as mixed numbers. Work on a separate sheet of paper.

25. $\frac{5}{2}$ **26.** $\frac{6}{5}$ **27.** $\frac{17}{3}$ **28.** $\frac{33}{4}$ **29.** $\frac{19}{8}$

Chapter Eight Additional Practice

Copy the problems below and solve them on a separate sheet of paper.

1. $\frac{2}{3} \times \frac{1}{3} =$

2. $\frac{3}{4} \times \frac{2}{3} =$

3. $\frac{4}{5} \times \frac{7}{8} =$

4. $\frac{3}{7} \div \frac{5}{6} =$

5. $\frac{3}{10} \div \frac{8}{9} =$

6. $\frac{2}{3} \div \frac{2}{5} =$

7. $\frac{4}{13} \div \frac{5}{8} =$

8. $\frac{5}{7} \times \frac{1}{2} =$

9. $9\frac{5}{8} \times 2\frac{3}{8} =$

10. $2\frac{1}{3} \times 8\frac{2}{5} =$

11. $7\frac{1}{2} \times 8\frac{1}{3} =$

12. $4\frac{2}{7} \times \frac{2}{3} =$

13. $4\frac{2}{3} \div 2\frac{1}{3} =$

14. $5\frac{6}{9} \div 2\frac{1}{3} =$

15. $8\frac{1}{4} \div 2\frac{1}{5} =$

16. $11\frac{2}{3} \div 4 =$

17. $\frac{3}{4} \div 15 =$

18. $\frac{3}{5} \times 10 =$

19. $5\frac{3}{5} \div 6 =$

20. $85 \div \frac{5}{6} =$

21. $4 \times 2\frac{1}{2} =$

22. $36 \div \frac{3}{8} =$

23. $3\frac{3}{4} \times 16 =$

24. $14 \times \frac{5}{7} =$

Solve the problems below on a separate sheet of paper.

Jane and Anne, the directors of Wildwood Day Camp, are planning a picnic at Lakeside Park.

1. Lakeside Park is $42\frac{1}{2}$ miles from the camp. The speed limit for buses on the road to the park is 35 miles per hour. How long will it take to get to the park?

2. There will be 28 people at the picnic. They will barbecue chicken parts. Each person will eat about $\frac{3}{4}$ of a pound of chicken. How many pounds of chicken must Jane buy?

3. One of the parents is sending $7\frac{1}{2}$ pounds of potato salad. How much potato salad will each person eat?

4. The camp has 6 thermos jugs. Each one holds $3\frac{4}{5}$ quarts. How much do they hold altogether?

5. Anne wants to take at least 24 quarts of liquid on the picnic. How much of this will not fit into the camp's thermos jugs?

Chapter Nine Additional Practice

Copy the problems below and solve them on a separate sheet of paper.

1. $\frac{5}{11} + \frac{2}{11}$

2. $\frac{7}{12} + \frac{9}{12}$

3. $\frac{2}{9} + \frac{2}{3}$

4. $\frac{4}{5} - \frac{3}{4}$

5. $\frac{7}{8} - \frac{1}{6}$

6. $5\frac{3}{8} - 7\frac{1}{8}$

7. $8\frac{2}{3} + 2\frac{4}{5}$

8. $9\frac{1}{8} + \frac{4}{7}$

9. $8\frac{4}{11} - 2\frac{3}{11}$

10. $9\frac{2}{3} - 5\frac{3}{5}$

11. $12\frac{3}{8} - 7\frac{4}{9}$

12. $8\frac{2}{10} - 7$

13. $5\frac{2}{3} + 7$

14. $8 - 2\frac{1}{4}$

15. $6\frac{2}{3} + 3\frac{1}{5}$

16. $12\frac{1}{4} - 7\frac{2}{3}$

Solve the problems below on a separate sheet of paper.

1. Sam and Cindy will hike from the picnic area to the lake along Frontier Trail. The trail is $5\frac{1}{4}$ miles long. Coming back, they will take the fire road. That road is $2\frac{7}{8}$ miles long. How much longer is Frontier Trail than the fire road?

2. Sam and Cindy hiked $5\frac{1}{4}$ miles to the lake. They hiked $2\frac{7}{8}$ miles along the fire road. After lunch they hiked along the Scenic Trail, which is $4\frac{1}{2}$ miles long. How many miles did they hike that day?

3. Jack bought a piece of carpet $7\frac{2}{5}$ yards long. He also bought one $9\frac{4}{7}$ yards long. How much carpet did he buy altogether?

4. Kitty had a piece of lumber $9\frac{1}{2}$ feet long. She used $7\frac{3}{4}$ feet. How much lumber was left?

5. The shelf in one closet is $6\frac{2}{3}$ feet long. The shelf in another closet is $8\frac{3}{8}$ feet long. How much longer is the second shelf?

Chapter Ten
Additional Practice

Solve the problems below on a separate sheet of paper.

1. $\begin{array}{r} 685.092 \\ 49.357 \\ +821.47 \\ \hline \end{array}$

2. $\begin{array}{r} 795.013 \\ -\ 62.252 \\ \hline \end{array}$

3. $\begin{array}{r} 4{,}282.36 \\ \times\ 357.15 \\ \hline \end{array}$

4. $\begin{array}{r} 862.373 \\ \times\quad 100 \\ \hline \end{array}$

5. $\begin{array}{r} 672.30 \\ \times 429.82 \\ \hline \end{array}$

6. $\begin{array}{r} 358.901 \\ -147.29 \\ \hline \end{array}$

7. $10\overline{)3.206}$

8. $100\overline{)54.92}$

9. $8\overline{)348.2}$

10. $6.2\overline{)105.40}$

11. $5.72\overline{)131.56}$

12. $4.63\overline{)7{,}892}$

Change these decimals to fractions on a separate sheet of paper. Reduce the fractions to lowest terms.

13. .88 **14.** .56 **15.** .007 **16.** .352 **17.** .165

Two of the three numbers in each group are equal. Write the number in each group that is *not* equal to the other two. Work on a separate sheet of paper.

18. .08 .008 $\frac{8}{100}$

19. .5 .50 $\frac{2}{5}$

20. .12 .121 $\frac{6}{50}$

21. .4 .04 $\frac{1}{25}$

22. .30 $\frac{1}{3}$ .3

23. .57 5.7 $\frac{57}{100}$

24. Kay drives to work and back every day. It is 13.7 miles each way. She works 5 days a week. How many miles does she drive each week? She works 50 weeks a year. How many miles does she drive in a year?

25. Ken had $231.56 in his checking account. He put in $92. Then he wrote a check for $181.23. How much was left in his checking account?

26. Mark bought three pairs of jeans that were on sale. The first pair cost $28.99. The second pair cost $35.99. The third pair cost $32.99. Find the average cost per pair of jeans.

Chapter Eleven Additional Practice

Solve the problems below on a separate sheet of paper.

1. What percent of 90 is 6?

2. What percent of 117 is 90?

3. 32% of 96 =

4. 23% of ______ = 8

5. 12% of 124 =

6. 59% of 82 =

7. 40% of ______ = 22

8. What % of 80 is 20?

9. 46% of 400 = ______

10. 55% of ______ = 330

11. 35% of ______ = 70

12. What % of 125 is 25?

Copy the chart below on a separate sheet of paper. Each row has one number in either the percent, decimal, or fraction column. Fill in the other two columns in each row. Work on a separate sheet of paper.

Percent	Decimal	Fraction
13. 8%		
14.	.75	
15.		$\frac{1}{4}$
16.	.23	
17.		$\frac{2}{5}$
18. 46%		
19.		$\frac{3}{4}$
20.	$.36\frac{1}{2}$	
21.		$4\frac{1}{4}$

Chapter Twelve Additional Practice

Find out if each pair of fractions below form a proportion. Cross multiply on a separate sheet of paper.

1. $\frac{12}{18}$ $\frac{2}{3}$

2. $\frac{3}{30}$ $\frac{4}{40}$

3. $\frac{12}{90}$ $\frac{3}{7}$

4. $\frac{1}{8}$ $\frac{2}{16}$

5. $\frac{7}{12}$ $\frac{5}{8}$

6. $\frac{1\frac{1}{4}}{6}$ $\frac{4}{3}$

Find the unknown term in each proportion below. Work on a separate sheet of paper.

7. $\frac{8}{?} = \frac{24}{36}$

8. $\frac{5}{7} = \frac{?}{21}$

9. $\frac{3}{8} = \frac{?}{32}$

10. $\frac{?}{30} = \frac{6}{15}$

11. $\frac{3}{18} = \frac{?}{6}$

12. $\frac{10}{?} = \frac{1}{5}$

13. $\frac{?}{5} = \frac{8}{20}$

14. $\frac{12}{?} = \frac{8}{12}$

15. $\frac{15}{45} = \frac{?}{3}$

16. $\frac{16}{8} = \frac{10}{?}$

17. $\frac{?}{100} = \frac{11}{25}$

18. $\frac{21}{?} = \frac{7}{40}$

Solve these problems on a separate sheet of paper.

1. Six apples cost $1.44. How much would four apples cost?
2. Three ears of corn is $.60. How much would five ears cost?
3. Eight cans of chili cost $4.80. How much would seven cans cost?
4. Four pounds of chicken cost $3.24. How much would three pounds cost?
5. Watermelon is on sale for $.29 per pound. Find the cost of a 15 pound watermelon.

Chapter Thirteen Additional Practice

Study the graph below. Answer the questions on a separate sheet of paper.

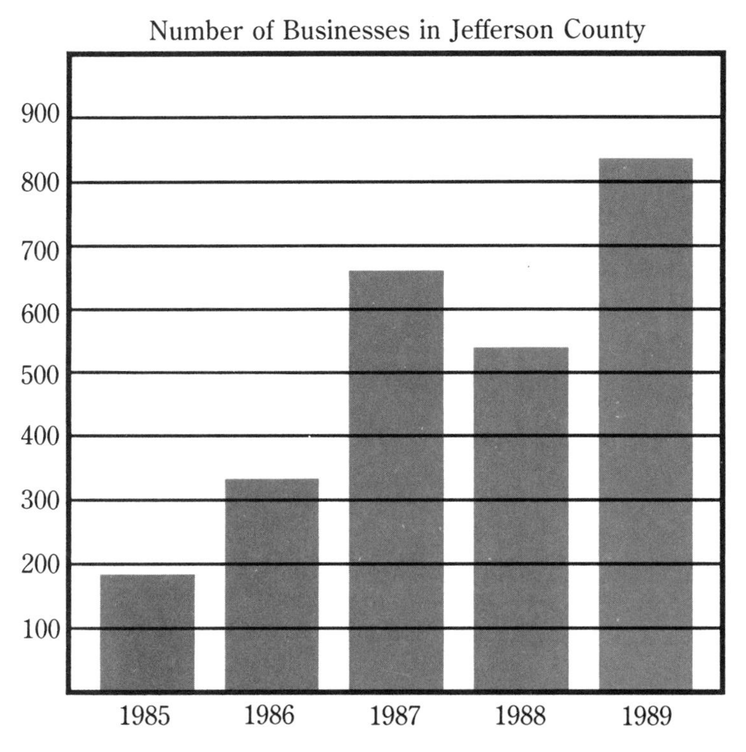

1. In what year did Jefferson County have the most businesses? About how many were there?
2. In what year did Jefferson County have the fewest businesses? About how many were there?
3. Between which two years did the numbers of businesses increase most? About how much was this increase?
4. What was the difference in the number of businesses between 1985 and 1989?
5. What is the range in the number of businesses from 1985 to 1989?

Chapter Fourteen Additional Practice

Read this group of numbers:

395 453 275 429 362

1. Find the range of the group of numbers above.
2. Find the median of the group of numbers above.
3. Find the mode of the group of numbers above.
4. Find the mean of the group of numbers above.

Read this group of numbers:

8,915 8,369 7,593 8,005 7,847 7,322

5. Find the range of the group of numbers above.
6. Find the median of the group of numbers above.
7. Find the mode of the group of numbers above.
8. Find the mean of the group of numbers above.
9. Make a frequency table for this group of numbers.

   29 33 23 30 20 27 30 33 29 27 29 29
   29 36 42 36 29 29 29 37 38 29 29 32

10. Use the frequency table to find the mode.
11. The list below shows the grades on a mathematics test for one class. Make a frequency table for the grades.

Allen, A.	72	Kaye, L.	75
Bastide, A.	84	Kim, D.	92
Browne, N.	68	Morgan, S.	87
Castorini, L.	84	Nogales, P.	79
Hernandez, M.	90	Owens, K.	84
Leary, P.	70	Patrocoles, C.	75

12. Use the frequency table to find the mode.
13. What is the range of the grades?
14. What is the mean grade?

Chapter Fifteen Additional Practice

Solve the problems below on a separate sheet of paper.

1. 2 yards = ______ inches
2. 4 quarts = ______ ounces
3. A backyard is 52 feet long and 24 feet wide. What is its perimeter?
4. A road is 39 miles long and 48 feet wide. What is its area?
5. A suitcase is 35 inches long, 23 inches wide, and 7 inches high. What is its volume?
6. Connie hiked for 3 hours. She went 8 miles. How many miles an hour did she hike?

Which of each pair of measurements is greater?

7. 79 inches or 2 yards
8. 2 gallons or 203 ounces
9. 500 feet or 50 yards

Which temperature of each pair is higher?

10. 41° F or 35° C
11. 15° C or 95° F
12. 68° F or 55° C
13. One room is 17 feet long and 14 feet wide. A second room is 20 feet long and 12 feet wide. Which room is larger? How much larger is it?
14. Ann's yard is 26 feet long and 35 feet wide. She wishes to plant a hedge around its edge. How many feet of hedge does she need?
15. A train traveled for 7 hours at 77 miles an hour. How far did it go?
16. Natalie drove 657 miles in 2 days. She drove the same distance each day. How far did she drive each day?
17. Paul drove 497 miles from Clarksville to Evanston. He drove at 53 miles an hour. How many hours did he drive?
18. Mike drove 389 miles in 6 hours. John drove 315 miles in 5 hours. Who drove faster? How much faster did he drive?

Chapter Sixteen Additional Practice

Change these units of measures on a separate sheet of paper.

1. Change 22 kilograms to grams.
2. Change 6,200 milligrams to grams.
3. Change 292 deciliters to liters.
4. Change 750 centigrams to grams.
5. Change 890 liters to dekaliters.
6. Change 68 liters to centiliters.
7. Change 8,397 meters to kilometers.
8. Change 5,734 grams to kilograms.
9. Change 14 hectoliters to liters.
10. Change 9 kilometers to meters.

Solve the problems below on a separate sheet of paper.

1. Norma wants to paint the baseboard around her room. The room is 8 meters long and 6 meters wide. How many meters must she paint?
2. Danny drove 589 kilometers in 6 hours. How many kilometers did he average in 1 hour?
3. Martin's company used 12 kilograms of paper during January. It used 9 kilograms during February. During March it used 11 kilograms. It used 13 kilograms in April and 9 kilograms in May. What was the average amount used in one month?
4. Terry drove 237 kilometers on Monday. She drove 182 kilometers on Tuesday. On Wednesday she drove 302 kilometers. She drove 97 kilometers on Thursday and 129 kilometers on Friday. What was the average number of kilometers she drove each day?
5. Terry used 48 gallons of gas for the week. About how many liters of gas is that?

Chapter Seventeen Additional Practice

Solve the problems below on a separate sheet of paper.

1. Find the perimeter of a square whose side is 34 centimeters.
2. Find the perimeter of a rectangle whose length is 8 feet and width is 13 feet.
3. Find the area of a square whose side is 79 inches.
4. Find the area of a rectangle whose length is 21 miles and width is 16 miles.
5. Find the area of a parallelogram whose base is 64 meters and whose height is 38 meters.
6. Find the area of a triangle whose base is 51 inches and whose height is 42 inches.
7. A triangle has angles of 64° and 38°. What is the size of the third angle?
8. Find the area of a triangle whose base is 54 inches and whose height is 39 inches.
9. A triangle has angles of 55° and 26°. What is the size of the third angle?
10. Find the square root of 169.
11. Find the area of the rectangle shown below.
12. Find the area of the triangle shown below.
13. Find the area of the parallelogram shown below.

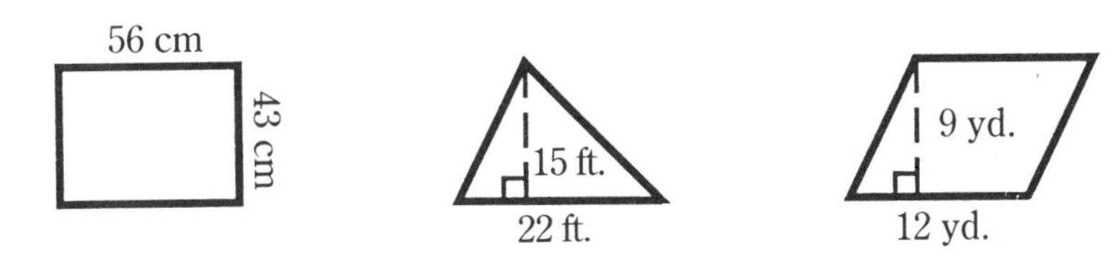

Chapter Eighteen Additional Practice

Solve the problems below on a separate sheet of paper.

1. Every circle has a radius and a diameter. Which is smaller?
2. What mixed decimal approximately represents π?
3. The diameter of a circle is 27 meters. Find the area.
4. The diameter of a circle is 39 inches. Find the circumference.
5. The radius of a circle is 9 feet. Find the area.
6. The radius of a circle is 16 yards. Find the circumference.
7. The diameter of a circle is 92 centimeters. Find the area.
8. The diameter of a circle is 43 meters. Find the area.
9. The radius of a circle is 18 inches. Find the area.
10. The radius of a circle is 32 yards. Find the area.

Study each group of words below. Three of the words in each group have something in common. One word does not belong. Number a separate sheet of paper and write down each word that doesn't belong.

11. radius   diameter   circumference   square
12. isosceles   parallelogram   scalene   equilateral
13. circle   rectangle   parallelogram   square
14. rectangle   square   quadrilateral   triangle

Copy the number of each picture below on a separate sheet of paper. Next to each number write the name of the correct shape from the box.

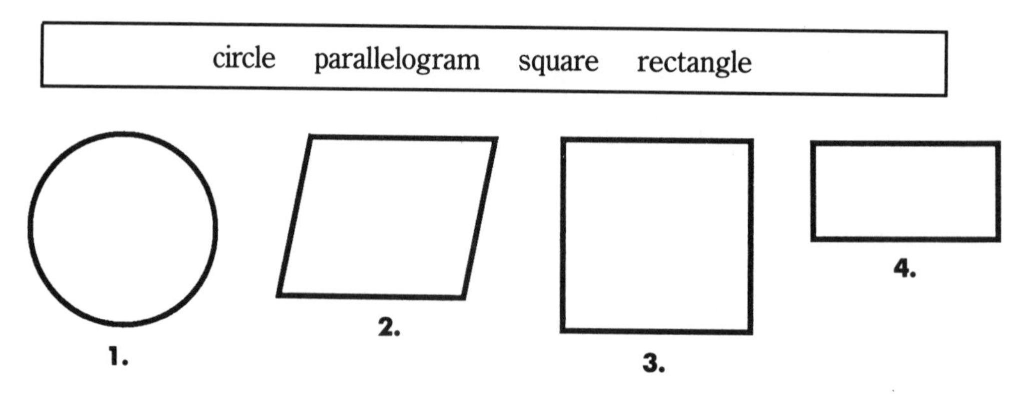

Chapter Nineteen Additional Practice

Solve these equations on a separate sheet of paper. Check your answers.

1. $x - 12 = 52$
2. $4(6 + 3) = y$
3. $57 - (6 \times 4) = d$
4. $80 = 50 + x$
5. $(7 + 5) - (6 - 4) = z$
6. $37 - (29 - 5) =$
7. $(18 \div 9) + (6 \times 9) =$
8. $9(6 - 7) =$
9. $v + 9 = 32$
10. $4^2 \times 3^3 = x$
11. $\frac{x}{28} = 168$
12. $8d + 17 = 81$
13. $15(6) - x = 67$
14. $7s - 9 = 54$
15. $3z = 84$
16. $6^5 - 4^3 = y$
17. $10m - 26 = 44$
18. $7(6) - n = 41$
19. $5x + 3 = 53$
20. $3x - 7 = 14$
21. $a - 26 = 47$
22. $15n = 225$
23. $6^2 - 4^2 = C$
24. $8(3 + 4) = n$
25. $11x = 132$
26. $z + 18 = 47$
27. $2x - 14 = 10$
28. $5^2 - 2^3 = a$
29. $7(4 + 6) = p$
30. $5y + 13 = 48$

Make up an equation to solve each problem below. Then solve the equations.

1. Jack and Margo typed 96 letters between them. Jack typed 3 times as many letters as Margo did. How many letters did each person type?
2. Mary and Carl have driven 240 miles between them. Mary has driven 4 times as many miles as Carl has. How many miles has each person driven?
3. Alex is one-third as old as his father. If Alex's father is now 36, how old is Alex?

Multiplication Table

Look at the multiplication chart below. It is easy to find the product of any two numbers on the chart. Find one number in the first column going down. Find the second number in the top row going across. Move across from the first column and down from the top row. The box in which the row and column meet contains the product of the two numbers.

1	2	3	4	5	6	7	8	9	10
2	4	6	8	10	12	14	16	18	20
3	6	9	12	15	18	21	24	27	30
4	8	12	16	20	24	28	32	36	40
5	10	15	20	25	30	35	40	45	50
6	12	18	24	30	36	42	48	54	60
7	14	21	28	35	42	49	56	63	70
8	16	24	32	40	48	56	64	72	80
9	18	27	36	45	54	63	72	81	90
10	20	30	40	50	60	70	80	90	100

Division Table

Look at the division chart below. It is easy to find the quotient of any two numbers on the chart. Find the divisor in the first column going down. Move across that row until you come to the dividend you want. Then move up that column until you get to the top row. The number in the box of the top row is the quotient.

1	2	3	4	5	6	7	8	9	10
2	4	6	8	10	12	14	16	18	20
3	6	9	12	15	18	21	24	27	30
4	8	12	16	20	24	28	32	36	40
5	10	15	20	25	30	35	40	45	50
6	12	18	24	30	36	42	48	54	60
7	14	21	28	35	42	49	56	63	70
8	16	24	32	40	48	56	64	72	80
9	18	27	36	45	54	63	72	81	90
10	20	30	40	50	60	70	80	90	100

Table of Roots and Square Roots

The table below can be used to find the square roots of whole numbers between 1 and 100.

The square roots in this chart are given to the nearest thousandth.

	Square	Square Root
1	1	1.000
2	4	1.414
3	9	1.732
4	16	2.000
5	25	2.236
6	36	2.449
7	49	2.646
8	64	2.828
9	81	3.000
10	100	3.162
11	121	3.317
12	144	3.464
13	169	3.606
14	196	3.742
15	225	3.873
16	256	4.000
17	289	4.123
18	324	4.243
19	361	4.359
20	400	4.472
21	441	4.583
22	484	4.690
23	529	4.796
24	576	4.899
25	625	5.000
26	676	5.099
27	729	5.196
28	784	5.292
29	841	5.385
30	900	5.477
31	961	5.568
32	1,024	5.657
33	1,089	5.745
34	1,156	5.831
35	1,225	5.916
36	1,296	6.000
37	1,369	6.083
38	1,444	6.164
39	1,521	6.245
40	1,600	6.325
41	1,681	6.403
42	1,764	6.481
43	1,849	6.557
44	1,936	6.633
45	2,025	6.708
46	2,116	6.782
47	2,209	6.856
48	2,304	6.928
49	2,401	7.000
50	2,500	7.071
51	2,601	7.141
52	2,704	7.211
53	2,809	7.280
54	2,916	7.348
55	3,025	7.416
56	3,136	7.483
57	3,249	7.550
58	3,364	7.616
59	3,481	7.681
60	3,600	7.746
61	3,721	7.810
62	3,844	7.874
63	3,969	7.937
64	4,096	8.000
65	4,225	8.062
66	4,356	8.124
67	4,489	8.185
68	4,624	8.246
69	4,761	8.307
70	4,900	8.367
71	5,041	8.426
72	5,184	8.485
73	5,329	8.544
74	5,476	8.602
75	5,625	8.660
76	5,776	8.718
77	5,929	8.775
78	6,084	8.832
79	6,241	8.888
80	6,400	8.944
81	6,561	9.000
82	6,724	9.055
83	6,889	9.110
84	7,056	9.165
85	7,225	9.220
86	7,396	9.274
87	7,569	9.327
88	7,744	9.381
89	7,921	9.434
90	8,100	9.487
91	8,281	9.539
92	8,464	9.592
93	8,649	9.644
94	8,836	9.695
95	9,025	9.747
96	9,216	9.798
97	9,409	9.849
98	9,604	9.899
99	9,801	9.950
100	10,000	10.000

Index

M

N

O

P

Q

R

S

T

V

W

Z